SPECTRA FOR THE IDENTIFICATION
OF MONOMERS IN FOOD PACKAGING

Commission of the European Communities
Directorate-General for Science, Research and Development
Community Bureau of Reference (BCR)

Spectra for the Identification of Monomers in Food Packaging

by

Jane Bush
*Maurice Palmer Associates,
Cambridge, United Kingdom*

John Gilbert
*Ministry of Agriculture, Fisheries & Food,
Norwich, United Kingdom*

and

Xabier Goenaga
*Community Bureau of Reference, DG XII, Commission of the European Communities,
Brussels, Belgium*

KLUWER ACADEMIC PUBLISHERS
DORDRECHT / BOSTON / LONDON

Library of Congress Cataloging-in-Publication Data

Bush, Jane.
    Spectra for the identification of monomers in food packaging / by
Jane Bush, John Gilbert, Xabier Goenaga.
        p.   cm.
    At head of title: Commission of the European Communities,
Directorate General for Science, Research, and Development,
Community Bureau of Reference (BCR)

    1. Food contamination--Handbooks, manuals, etc. 2. Monomers-
-Spectra--Handbooks, manuals, etc. 3. Plastics in packaging-
-Handbooks, manuals, etc.   I. Gilbert, John. II. Goenaga, Xabier.
III. Commission of the European Communities. Community Bureau of
Reference. IV. Title.
TX271.P63887 1993
363.19'2--dc20
                                                    93-25450

ISBN 978-90-481-4304-7

Publication arrangements by
Commission of the European Communities
Dissemination of Scientific and Technical Knowledge Unit,
Directorate-General Information Technologies and Industries and Telecommunications, Luxembourg

EUR 14515EN
© 1993 ECSC, EEC, EAEC, Brussels and Luxembourg
Softcover reprint of the hardcover 1st edition 1993

LEGAL NOTICE
Neither the Commission of the European Communities nor any person acting on behalf of
the Commission is responsible for the use which might be made of the following information.

Published by Kluwer Academic Publishers,
P.O. Box 17, 3300 AA Dordrecht, The Netherlands.

Kluwer Academic Publishers incorporates the publishing programmes of
D. Reidel, Martinus Nijhoff, Dr W. Junk and MTP Press.

Sold and distributed in the U.S.A. and Canada
by Kluwer Academic Publishers,
101 Philip Drive, Norwell, MA 02061, U.S.A.

In all other countries, sold and distributed
by Kluwer Academic Publishers Group,
P.O. Box 322, 3300 AH Dordrecht, The Netherlands.

Printed on acid-free paper

CONTENTS

vi

Substance	CAS No.	PM/REF No.	Mol. wgt.	Page No.
Isobutene	00115-11-7	19000	56.11	238
Lauric acid	00143-07-7	19470	200.32	239
Lignocellulose	11132-73-3	19510	–	no entry
Maleic acid	00110-16-7	19540	116.07	243
Maleic anhydride	00108-31-6	19960	98.06	245
Methacrylic acid	00079-41-4	20020	86.09	248
Methacrylic acid, butyl ester	00097-88-1	20110	142.20	251
Methacrylic acid, ethyl ester	00097-63-2	20890	114.15	254
Methacrylic acid, ethylene glycol monoester	00868-77-9	21190	130.14	257
Methacrylic acid, isobutyl ester	00097-86-9	21010	142.20	260
Methacrylic acid, isopropyl ester	04655-34-9	21100	128.17	263
Methacrylic acid, methyl ester	00080-62-6	21130	100.12	264
Methacrylic acid, propyl ester	02210-28-8	21340	128.17	267
Methacrylic acid, sec-butyl ester	02998-18-7	20140	142.20	268
Methacrylic acid, tert-butyl ester	00585-07-9	20170	142.20	269
Methacrylic anhydride	00760-93-0	21460	154.17	270
Methacrylonitrile	00126-98-7	21490	67.09	273
Methanol	00067-56-1	21550	32.04	277
N-Methylolacrylamide	00924-42-5	21940	101.11	280
4-Methyl-1-pentene	00691-37-2	22150	84.16	281
Myristic acid	00544-63-8	22350	228.38	284
1,5-Naphthalene diisocyanate	03173-72-6	22420	210.19	288
Nitrocellulose	09004-70-0	22450	–	292
1-Nonanol	00143-08-8	22480	144.26	293
Octadecyl isocyanate	00112-96-9	22570	295.51	296
1-Octanol	00111-87-5	22600	130.23	300
1-Octene	00111-66-0	22660	112.22	303
Palmitic acid	00057-10-3	22780	256.43	307
Pentaerythritol	00115-77-5	22840	136.15	311
1-Pentanol	00071-41-0	22870	88.15	314
Phenol	00108-95-2	22960	94.11	317
1,3-Phenylenediamine	00108-45-2	23050	108.14	320
Phenyl isocyanate	00103-71-9	23125	119.12	323
Phosphoric acid	07664-38-2	23170	98.00	327
Phthalic acid, diallyl ester	00131-17-9	23230	246.27	329
Phthalic anhydride	00085-44-9	23380	148.12	332
alpha-Pinene	00080-56-8	23470	136.23	335
beta-Pinene	00127-91-3	23500	136.23	336
Polyethyleneglycol	25322-68-3	23590	–	339
Polypropyleneglycol	25322-69-4	23650	–	340
1,2-Propanediol	00057-55-6	23740	76.12	342
1-Propanol	00071-23-8	23800	60.10	345
2-Propanol	00067-63-0	23830	60.10	348
Propionaldehyde	00123-38-6	23860	58.08	351
Propionic acid	00079-09-4	23890	74.08	352
Propionic anhydride	00123-62-6	23950	130.14	356
Propylene	00115-07-1	23980	42.08	359
Propylene oxide	00075-56-9	24010	58.08	360
Resin acids and resin acids	73138-82-6	24070	–	362

CHAPTER 1

Legal Framework

In the European Community one of the main controls on materials
and articles for food contact use is through Directive
90/128/EEC (1) which restricts the range of monomers and other
starting substances that can be used for the production of
plastics. The control is through the use of a positive list of
authorised substances grouped into section A of approved
substances and section B of substances with provisional approval
pending a decision on inclusion in section A. Directive 90/128
together with subsequent amendments listed 163 substances in
section A, although it was envisaged that the situation would
never be static and there would always be movements from section
B to section A as well as approval and adoption of new
substances. In addition to positive list controls the Directive
places limits on certain substances in the form of maximum
levels of residual substance permitted in the finished plastic
(QM limit) and/or maximum amount of substance permitted to
migrate into foods or food simulants under defined conditions
(SML limit).

Implementation of legislation

Although the principles that are intended to govern the control
of materials and articles are clear from the Directive, the
practical problems of implementation or the development of
approaches that should be adopted by enforcement authorities in
real situations have not yet been addressed. The most
systematic approach to control has been elaborated in the
Netherlands (2) to meet Dutch Regulations in existence before EC
90/128. The approach used has been to initially identify
polymeric materials by infra-red spectroscopy and then to
identify substances in solvent extracts by gas
chromatography/mass spectrometry and liquid chromatography.

2

Practical applications of this approach over many years has
shown that considerable experience is required in knowing what
polymer is used in what food contact situation, as well as the
likely additives and other constituents that might be
anticipated. Spectroscopic and chromatographic data-base
information as well as access to authentic reference standards
is essential to carry out this enforcement work.

In 1990, the UK Ministry of Agriculture, Fisheries and Food
funded a project with Maurice Palmer Associates in Cambridge
(UK) aimed at collecting monomers and other starting substances
from industry to form a reference collection for use by
enforcement authorities. It was found that often it was
difficult to identify a source of such substances. Further
support was given in 1991 by the Community Bureau of Reference
(BCR) of the Commission of European Communities for the
completion of the reference collection, the preparation of a
database of physicochemical data and a handbook of spectra of
monomers and starting substances. This Handbook is seen as
being primarily of practical assistance to enforcement
laboratories in the European Community but may well also be
useful to industry, Universities and other research
establishments. The book is envisaged as being a reference
source of information that can be used by relatively
inexperienced laboratories to develop expertise in the area of
food contact materials. In addition to the spectroscopic and
other information contained in the Handbook, reasonable requests
for the substances themselves will be met free of charge on
application to the address below. The substances will be
supplied as reference standards (1 g) or as reference solutions
where appropriate from:-

Plastics Reference Collection
Ministry of Agriculture, Fisheries and Food,
Food Science Laboratory,
Norwich Research Park,
Colney,
Norwich, NR4 7UQ (U.K.)

Scope of the Handbook.

Altogether 106 of the 163 substances listed in section A have been obtained as commercially available samples either from industry or from laboratory chemical suppliers. Certain substances which were natural materials of ill-defined composition such as rubber, cellulose and albumin were not included in the collection although they are listed in the Handbook. Volatile monomers and gases (17 in total) which could not be easily retained in pure form were prepared in solution at a defined concentration, the solvent being chosen on the basis of that most likely to be suitable for the purposes of analysis. It was appreciated that with the constant amendments to Directive 90/128 EEC no Handbook could ever be entirely complete or up-to-date and this must be appreciated as an inevitable limitation. An extension to this Handbook will be needed to include those additional monomers and starting substances which will have been granted approval for use.

Entries for each substance give the structural formula, the CAS number and the PM reference number which is the number by which substances are listed in Directive 90/128. Alternative names whether systematic IUPAC nomenclature or trivial names are given as a further aid to identification. Physical characteristics such as solubility in common organic solvents are described. Some indications for handling are given, but safety requirements vary from country to country and therefore indications of safety given are very general and not intended to have any legal status. For example, in the UK safe handling of chemicals is regulated by the Control of Substances Hazardous to Health Regulations (COSHH) and the indications in this Handbook are not meant to surpass the requirements of making a full COSHH assessment. Information provided on stability is intended as advisory, being based on that provided in good faith by suppliers or from published sources, and cannot be taken as definitive.

4

Information on current uses of monomers and food contact
applications, whilst invaluable to the enforcement laboratory,
is often anecdotal and thus difficult to assemble in any
systematic way. There is also the added complication that
whilst information particularly on applications may be relevant
for one country there may be quite different practice in other
countries. To cope with these difficulties entries on usage and
applications for each substance have been circulated widely for
comment in the UK, The Netherlands, Germany, France and Spain as
well as to the Association of Polymer Manufacturers Europe
(APME).

The infra-red spectra are intended to complement the already
available library collections of Hummel and Scholl (3) and
Sadtler (4). Spectra are given for substances already in these
libraries as well as for substances for which spectra have not
been previously published. All spectra have been run on more
than one instrument as a check on the authenticity of the data
and frequently in a number of different formats. In a small
number of instances where inconsistencies were found in spectra
either between those for the Handbook and other library spectra
or between those obtained on two different instruments, unless
the differences could be explained on interpretation then
additional samples were obtained from a second source. As the
infra-red spectra were envisaged as likely to be used primarily
as an aid for identification of polymers, in a number of
instances for major polymers the spectra of the polymers are
also given.

Mass spectra are similarly meant to complement existing data
bases, and have also been obtained on two or more instruments in
order to control and cross-check the quality of spectra
ultimately adopted for the Handbook. In some instances through
lack of volatility or for other reasons mass spectra could not
be obtained - where this was the case it is indicated, although
where suitable spectra could be obtained of volatile
derivatives, then this approach was adopted. Gas
chromatographic retention indices are provided as an additional
aid to assist in identification of monomers.

The Handbook also contains a brief entry on analytical methods which gives an outline of the approach to be used for measuring to test for compliance with QM or SML limits. Other bodies are actively engaged on method development and validation of analytical methods in support of 90/128 EEC, in particular the European Committee for Standardisation (CEN). Where it was known that a CEN method was in development this has been indicated together with a note of the organisation responsible for the method development. In order to meet the urgent need for more methods, the BCR Programme has undertaken a project for the development of methods for more than 30 substances with restrictions in 90/128/EEC. When ready these methods will be normalised by CEN. The Council of Europe has also been active in the area of assessment of published methods for the analysis of monomers and other starting substances, and ILSI Europe has developed an analytical methods data-base that gives a comprehensive coverage of published analytical procedures (5). In this Handbook we have cited what we believe to be the most relevant references to analytical methods in the published literature taking account of the likely availability of the technique employed and the relevance of the matrices tested.

It is intended that this Handbook should be used as a general reference source to assist in the identification of plastics materials and articles obtained for enforcement purposes. Infra-red and mass spectra for unknown materials can be compared with the reference spectra in the Handbook for identification and where appropriate the references to further analytical methods can be pursued.

6

Acknowledgement

We are very grateful to the laboratories and individuals listed
below who provided infra-red spectra and gas
chromatographic/mass spectral data:-

Mrs A. López de Sá and Mrs J. Gónzalez Gutierrez
CICC
Instituto Nacional del Consumo
Ministerio de Sanidad y Consumo
Madrid (Spain)

Dr D. Lopez-Delgado
Instituto de Materiales (sede D),
CSIC, Madrid (Spain)

Dr J.B.H. van Lierop
Food Inspection Service,
Nijenoord 6,
3552 AS Utrecht (Netherlands)

The success in producing this Handbook has also been due to the
assistance in supplying samples and to the information made
freely available from industrial and Government sources too
numerous to mention. We are especially grateful to Mr R. Ashby
(ICI, Wilton), Dr R. Franz (Fraunhofer Institute, Munich), Dr
P.Tice (PIRA, Leatherhead), Mr R. Rijk (TNO, Zeist), Dr L. Rossi
(EC, Brussels) and Mr D Shorten (BP Chemicals, UK) for their
advice and comments on entries to the Handbook.

References

(1) Commission Directive of 23 Feb 1990 relating to plastics
materials and articles intended to come into contact with
foodstuffs (90/128/EEC). Official Journal of the European
Communities No L 349/26, 1990.

(2) van Battum, D., and van Lierop, J.B.H., 1988, Food Additives
and Contaminants, 5, 381-395.

(3) D.O. Hummell and F. Scholl, Atlas of polymer and plastics analysis. Vol 1-3. Hanser publishers (Munich, Vienna, New York and Barcelona) and VCH (Weinheim, New York, Basel and Cambridge).

(4) Sadtler Monomers and Polymers Library 10600 spectra. Heyden & Son Ltd. (London).

(5) ILSI Data base of Analytical methods (ILSI Europe, Avenue E Mounier 83, Brussels)

Jane Bush (MPA Associates, Cambridge)
John Gilbert (MAFF, Norwich)
Xabier Goenaga (BCR, CEE, Brussels)

CHAPTER 2

Selection of entries

The monomers selected for entry are those in list A of
directive 90/128/EEC that have SML or QM restrictions. Since
it is the intention to update this directive annually, with
additions, deletions, and movement of substances from list B
to list A, readers should check the most recent amendment to
establish the exact status of a substance at the time of
enquiry.

Some substances listed in the directive could not be traced
and were not available either from industry or from the
normal chemical supply companies. In these cases therefore,
no reference substance is available. Further, in some cases
it was not possible to attribute a final use to some of
these substances. In these cases, no entry has been included
in the atlas except for a listing in the index.

Literature references

The selection criteria for references to analytical methods
were that the methods should relate to foods, simulants or
food contact materials. At the time of writing, analytical
methods are under development for all substances with
restrictions, under the auspices of CEN (Working Group 5 of
Technical Committee 194) and the BCR (DGXII/C/5,
Measurements and Testing Programme 1992). It is the
intention of DGIII/C/1 to provide a compendium of analytical
methods as part of the general "Notes for guidance" that
will complement directives. This compendium will include
methods that are finalised as well as a brief description of
the basis of methods under development. The current draft of
this compendium document is entitled "Guidelines for the
compliance with and enforcement of community directives on

materials and articles intended to come into contact with
foodstuffs: Sampling and methods of analysis".

Mass spectra

Mass spectra have been recorded under standardised
conditions and a common format for data presentation has
been adopted for each entry. Spectra were obtained using a
Finnigan Mat SSQ 70 mass spectrometer coupled to a Varian
3400 gas chromatograph. Ionisation was by electron impact at
70 eV with a source temperature of 150°C. The scanning range
was m/z 25-400. The GC was equipped with a J & W DB5
capillary column of length 30 m, internal diameter 0.32 mm
and film thickness of 0.52 um. To ensure the spectra in this
atlas are representative, mass spectra were also obtained in
a second laboratory using different instrumentation and
conditions. Finally, mass spectra were compared with library
reference spectra where available. In the large majority of
cases, agreement was good and a single representative
spectrum is presented for each entry. It should be
remembered however, that there can be differences in
fragmentation patterns between the common mass spectrometers
- magnetic sector, quadropole, mass selective detectors and
ion-traps, depending on in part, the operating conditions.

The 20 most intense ions are reported for the mass spectra
together with the molecular ion if observed. Where
substances could not readily be analysed by GC-MS (organic
acids for example) methyl esters were prepared and these
spectra reported. The rationale for this is that it is in
this chemical form that these substances would most commonly
be encountered in the analysis of food contact materials.
Included in the information for each entry is the relative
retention time for the substance compared to eicosane (C-20
alkane). This value is given only as an indication of the
chromatographic behaviour of the substance in question and
it is not a retention index. The GC temperature programme
used was: 3 min isothermally at 50°C then rising at 20°C/min

to 300°C and held for a further 20 min. Kovacs indices have been included for some entries.

Infra red spectra

The manner in which infra red spectra were obtained depended on the physical nature of the substance. The techniques used were formulation into KBr pellets and deposition onto KBr, KRS-5 or NaCl windows. Spectra were acquired using a Brucker FT model IFS 85 spectrophotometer linked to an Aspect 3000 computer. The scan range used was dependant on the techniques employed. The sample and reference cell were scanned 60 times at 1.4 sec/scan and the mean difference spectrum plotted. As with mass spectra, IR spectra were also obtained in a second laboratory using different instrumentation and conditions to ensure the spectra are representative. Finally, IR spectra were compared with library reference spectra where these were available.

Approximately 20 peaks have been selected for inclusion in the listed peak table. IR spectra for the most commonly used polymers have also been included in the atlas to assist in the identification of food contact plastics.

Safety

Physical characteristics, handling precautions and safety precautions are included in this handbook as a guide only. They are not intended to be complete or definitive. Readers must refer to National safety standards on these matters and take all appropriate precautions.

Abietic acid

CAS No.	– 00514–10–3
PM Ref. No.	– 10030
Restrictions	– none
Formula	– $C_{20} H_{30} O_2$
Molecular weight	– 302.46
Alternative names	– Sylvic acid.

Physical Characteristics   – Yellow/orange crystals, mp 139–142°C, bp 250°C. Soluble in alcohol, benzene, chloroform, ether and acetone.

Handling   – Store at room temperature (25°C).

Safety   – Irritant.

Availability   – Standard sample supplied.

Current uses   – Manufacture of methyl, vinyl and glyceryl esters.

Applications   – Used in lacquers and varnishes.

Methods of Characterisation – IR

         Mass Spectroscopy

Purity   – >90%

Abietic acid

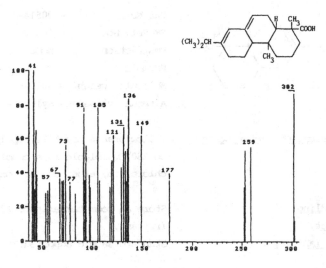

M/Z	Ion Intensity(%)	M/Z	Ion Intensity(%)
41	100.0	121	58.8
43	64.8	131	66.9
57	33.9	133	53.0
67	36.3	135	54.3
73	52.7	136	79.5
77	32.1	149	62.0
91	75.1	177	35.5
93	55.8	253	53.3
105	73.2	259	55.2
119	47.4	302	87.4

Spectrometer   :Finnigan Mat SSQ 70
Inlet System   :Capillary GC/MS
Source Temperature:150°C
Electron Energy   :70 eV
Scan Range   :25-400

Abietic acid — Transmission Infra Red

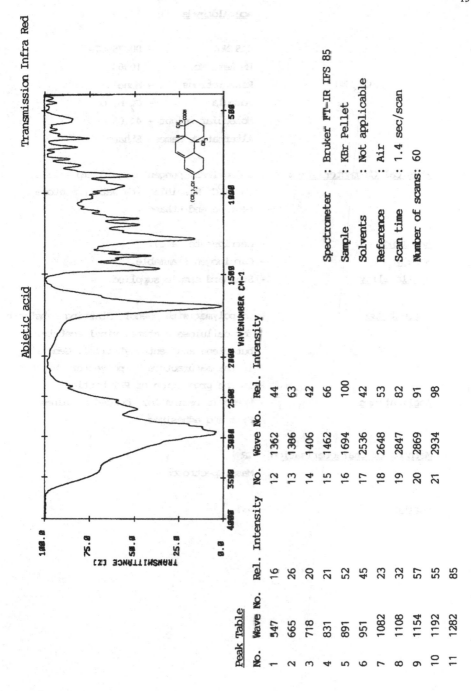

Spectrometer	: Bruker FT-IR IFS 85
Sample	: KBr Pellet
Solvents	: Not applicable
Reference	: Air
Scan time	: 1.4 sec/scan
Number of scans: 60	

Peak Table

No.	Wave No.	Rel. Intensity	No.	Wave No.	Rel. Intensity
1	547	16	12	1362	44
2	665	26	13	1386	63
3	718	20	14	1406	42
4	831	21	15	1462	66
5	891	52	16	1694	100
6	951	45	17	2536	42
7	1082	23	18	2648	53
8	1108	32	19	2847	82
9	1154	57	20	2869	91
10	1192	55	21	2934	98
11	1282	85			

Acetaldehyde

CH_3CHO

CAS No. - 00075-07-0
PM Ref. No. - 10060
Restrictions - none
Formula - C_2H_4O
Molecular weight - 44.05
Alternative names- Ethanal.

Physical Characteristics - Colourless, pungent liquid, mp -125°C, bp 21°C. Miscible with water, acetone, benzene and ethanol.

Handling - Refrigerate (4°C).
Safety - Carcinogen/Flammable.
Availability - Standard sample supplied.

Current uses - Co-polymer with phenol. Starting substance for cellulose acetate, vinyl acetate, butadiene and pentaerythritol. Generated in the manufacture of polyester chip used in the production of PET bottles.
Applications - Phenolic resins for moulded articles. Film and adhesives.

Methods of Characterisation - IR
Mass Spectroscopy

Purity - 99%

Acetaldehyde

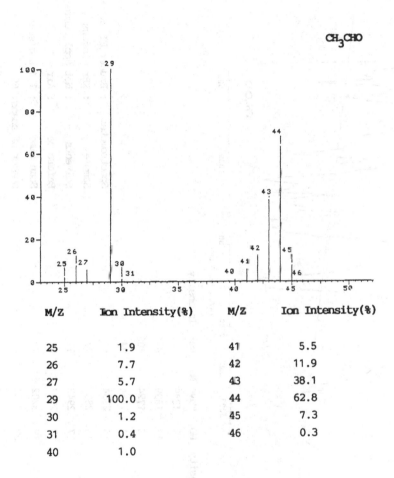

CH$_3$CHO

M/Z	Ion Intensity(%)	M/Z	Ion Intensity(%)
25	1.9	41	5.5
26	7.7	42	11.9
27	5.7	43	38.1
29	100.0	44	62.8
30	1.2	45	7.3
31	0.4	46	0.3
40	1.0		

Spectrometer	:Finnigan Mat SSQ 70
Inlet System	:Capillary GC/MS
Source Temperature	:150°C
Electron Energy	:70 eV
Scan Range	:25-400

16

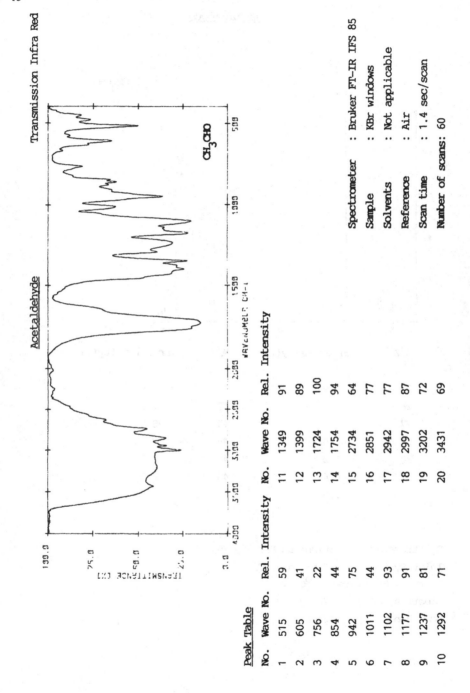

Acetaldehyde   Transmission Infra Red

CH₃CHO

Spectrometer   : Bruker FT-IR IFS 85
Sample   : KBr windows
Solvents   : Not applicable
Reference   : Air
Scan time   : 1.4 sec/scan
Number of scans: 60

Peak Table

No.	Wave No.	Rel. Intensity	No.	Wave No.	Rel. Intensity
1	515	59	11	1349	91
2	605	41	12	1399	89
3	756	22	13	1724	100
4	854	44	14	1754	94
5	942	75	15	2734	64
6	1011	44	16	2851	77
7	1102	93	17	2942	77
8	1177	91	18	2997	87
9	1237	81	19	3202	72
10	1292	71	20	3431	69

Acetic acid

CAS No.	– 00064–19–7
PM Ref. No.	– 10090
Restrictions	– none
Formula	– $C_2 H_4 O_2$
Molecular weight	– 60.05
Alternative names	– Ethanoic acid.

CH_3–COOH

Physical Characteristics – Colourless, pungent liquid, mp 16.6^{O}C, bp 118°C. Miscible with water, alcohol, glycerine & ether.

Handling – Store at room temperature (25°C).

Safety – Corrosive/Irritant/Flammable.

Availability – Standard sample supplied. 25% w/w solution with water.

Current uses – Modifier for cellulose. Melamine-formaldehyde co-polymers and acrylic adhesives. Preparation of polyetherimides. Polymerisation inhibitor for PET. In ethylene polymers, and polymers of adipic acid with diethylene glycol.

Applications – Films, fresh product blisterpacks, transparent windows in carton boxes.

Methods of Characterisation – IR
Mass Spectroscopy

Purity – Technical grade supplied by industry. Purity not declared.

18

Acetic acid

CH$_3$-COOH

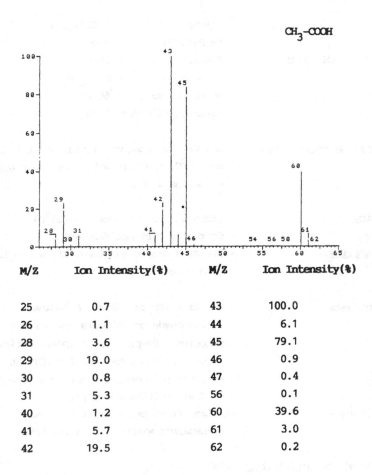

M/Z	Ion Intensity(%)	M/Z	Ion Intensity(%)
25	0.7	43	100.0
26	1.1	44	6.1
28	3.6	45	79.1
29	19.0	46	0.9
30	0.8	47	0.4
31	5.3	56	0.1
40	1.2	60	39.6
41	5.7	61	3.0
42	19.5	62	0.2

Spectrometer   :Finnigan Mat SSQ 70
Inlet System   :Capillary GC/MS
Source Temperature:150°C
Electron Energy   :70 eV
Scan Range   :25-400

Acetic acid, methyl ester

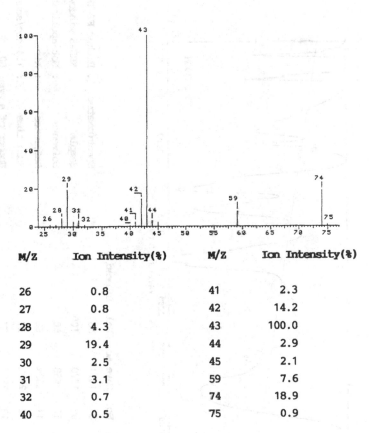

M/Z	Ion Intensity(%)	M/Z	Ion Intensity(%)
26	0.8	41	2.3
27	0.8	42	14.2
28	4.3	43	100.0
29	19.4	44	2.9
30	2.5	45	2.1
31	3.1	59	7.6
32	0.7	74	18.9
40	0.5	75	0.9

Spectrometer   :Finnigan Mat SSQ 70
Inlet System   :Capillary GC/MS
Source Temperature:150°C
Electron Energy   :70 eV
Scan Range   :25-400

Acetic acid

Transmission Infra Red

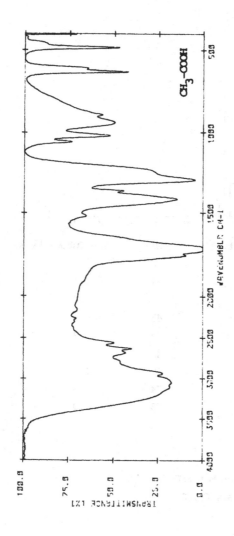

CH$_3$-COOH

Spectrometer : Bruker FT-IR IFS 85
Sample   : KRS-5 windows
Solvents   : Not applicable
Reference   : Air
Scan time   : 1.4 sec/scan
Number of scans: 60

Peak Table

No.	Wave No.	Rel. Intensity
1	481	53
2	629	58
3	891	43
4	937	50
5	1015	48
6	1053	27
7	1295	95

No.	Wave No.	Rel. Intensity
8	1414	85
9	1720	100
10	2558	50
11	2632	60
12	2682	57
13	2936	77
14	3094	83

Acetic acid, vinyl ester

$CH_3COOCH=CH_2$

CAS No.   – 00108-05-4
PM Ref. No.   – 10120
Restrictions   – SML=12 mg/kg
Formula   – $C_4 H_6 O_2$
Molecular weight   – 86.09
Alternative names   – Vinyl acetate.

Physical Characteristics – Colourless liquid, mp –100.2°C, bp 72.5°C. Sparingly soluble in water. Inhibited with 8-12 mg/kg hydroquinone monomethyl ether.

Handling – Store at room temperature (25°C).

Safety – Flammable/Irritant.

Availability – Standard sample supplied.

Current uses – Styrene polymers. Polyester resins. Starting substance for polyvinyl acetate and polyvinyl alcohol. Ethylene co-polymers (EVA, EVOH). Co-polymers with vinyl chloride.

Applications – Films, water soluble adhesives, dual ovenable trays. Stretch film, shrink film, barrier films. Coating paperboard for packaging ice-cream. Refrigerator trays.

Methods of Characterisation – IR
Mass Spectroscopy

Purity – 99.9%

Analytical methods – Capillary headspace GC with flame ionisation detection directly heating food simulants.

References – Method under development (PIRA, Leatherhead, UK).

Acetic acid, vinyl ester

$CH_3COOCH=CH_2$

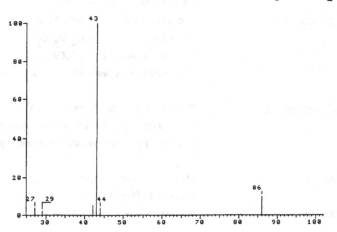

M/Z	Ion Intensity(%)
27	3.6
29	2.0
42	5.1
43	100.0
44	3.6
86	9.8

Spectrometer   :Finnigan Mat SSQ 70
Inlet System   :Capillary GC/MS
Source Temperature :150°C
Electron Energy   :70 eV
Scan Range   :25-400

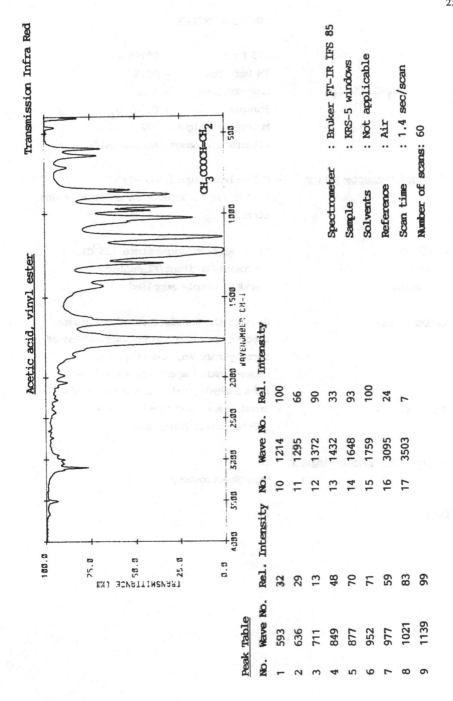

Acetic acid, vinyl ester

Transmission Infra Red

$CH_3COOCH=CH_2$

WAVENUMBER CM-i

Spectrometer : Bruker FT-IR IFS 85
Sample : KRS-5 windows
Solvents : Not applicable
Reference : Air
Scan time : 1.4 sec/scan
Number of scans: 60

Peak Table

No.	Wave No.	Rel. Intensity	No.	Wave No.	Rel. Intensity
1	593	32	10	1214	100
2	636	29	11	1295	66
3	711	13	12	1372	90
4	849	48	13	1432	33
5	877	70	14	1648	93
6	952	71	15	1759	100
7	977	59	16	3095	24
8	1021	83	17	3503	7
9	1139	99			

Acetic anhydride

CH$_3$-COOCOCH$_3$

CAS No.	- 00108-24-7
PM Ref. No.	- 10150
Restrictions	- none
Formula	- C$_4$ H$_6$ O$_3$
Molecular weight	- 102.09
Alternative names	- Acetic oxide.

Physical Characteristics — Colourless liquid, mp -73°C,
bp 138-140°C. Soluble in chloroform and
ether.

Handling — Store at room temperature (25°C).
Safety — Corrosive/Irritant/Flammable.
Availability — Standard sample supplied.

Current uses — Used in the manufacture of cellulose
acetate. Used in the polymerisation of
tetrahydrofuran, and ethylene.
Cross-linking agent for melamine-
formaldehyde resins and acrylic adhesives.
Applications — Bread wraps, and fresh produce
blisterpacks. Adhesives.

Methods of Characterisation — IR
Mass Spectroscopy

Purity — 99%

Acetic anhydride

$$CH_3-COOCOCH_3$$

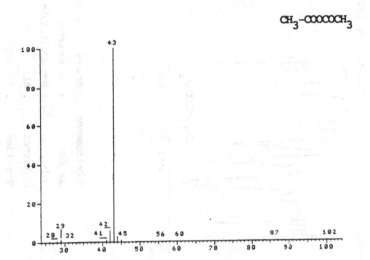

M/Z	Ion Intensity(%)	M/Z	Ion Intensity(%)
26	0.2	41	1.0
27	0.1	42	6.0
28	0.6	43	100.0
29	1.2	44	2.6
31	0.1	45	1.0
32	0.1	60	0.1
40	0.1	87	0.3

Spectrometer   :Finnigan Mat SSQ 70
Inlet System   :Capillary GC/MS
Source Temperature :150°C
Electron Energy   :70 eV
Scan Range   :25-400

26

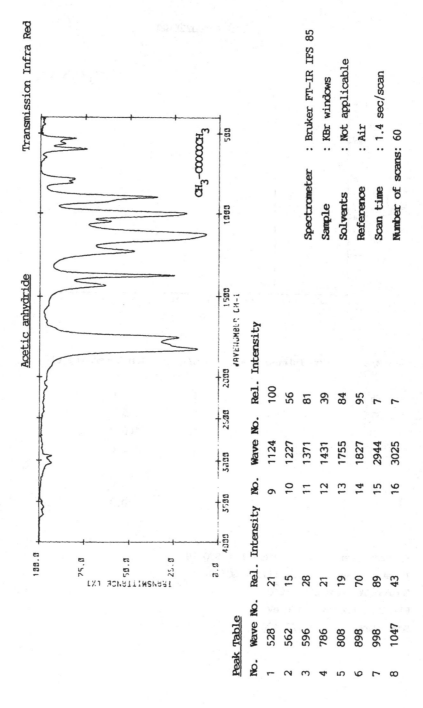

Acetic anhydride

Transmission Infra Red

$CH_3-COOOCH_3$

Spectrometer	:	Bruker FT-IR IFS 85
Sample	:	KBr windows
Solvents	:	Not applicable
Reference	:	Air
Scan time	:	1.4 sec/scan
Number of scans:		60

Peak Table

No.	Wave No.	Rel. Intensity	No.	Wave No.	Rel. Intensity
1	528	21	9	1124	100
2	562	15	10	1227	56
3	596	28	11	1371	81
4	786	21	12	1431	39
5	808	19	13	1755	84
6	898	70	14	1827	95
7	998	89	15	2944	7
8	1047	43	16	3025	7

Acetylene

CAS No.	– 00074–86–2
PM Ref. No.	– 10210
Restrictions	– none
Formula	– C_2H_2
Molecular weight	– 26.04
Alternative names	– Ethyne.

CH≡CH

Physical Characteristics – Colourless gas, mp -81°C (sublimes). Soluble in water, ether, and benzene.

Safety – Toxic.

Availability – No sample supplied.

Current uses – Polyacetylene. Starting substance in the synthesis of N-vinyl-2-pyrrolidinone, acrylates, acrylonitrile, acetaldehyde and vinyl monomers. Isoprene synthesis. Used in polyolefin synthesis.

Applications – A conductive polymer used as storage trays and racks. Poly(N-vinyl-2-pyrrolidinone) used to clarify beers and wines, also increases cross-linking rates. Synthetic rubbers. Coatings.

Acrylamide

$CH_2=CH-CONH_2$

CAS No.	– 00079–06–1
PM Ref. No.	– 10630
Restrictions	– SML(T)= not detectable (DL= 0.01 mg/kg)
Formula	– $C_3 H_5 N O$
Molecular weight	– 71.08
Alternative names	– 2–Propenamide; acrylic acid amide.

Physical Characteristics – Grey crystalline flakes, mp 84.5^{O}C, bp 125°C/25mmHg. Soluble in methanol and chloroform.

Handling – Refrigerate (4°C). Light sensitive.

Safety – Toxic/Irritant.

Availability – Standard sample supplied.

Current uses – Polyacrylamide gels. Cross–linking agent for styrene based polyester resins. Co–polymer with vinylidene chloride. Used in polyacrylates.

Applications – Flocculants in the production of sugars, for hydroponic growth media, coagulant used in water treatment.

Methods of Characterisation – IR
Mass Spectroscopy

Purity – 99%

Analytical methods –Simulants are treated directly with bromine reagent then analysed by capillary GC with nitrogen–selective detection. 2,3–dibromo–2–dimethylpropioamide is used as an internal standard. Food samples require additional silica–gel cartridge clean-up step. Confirmation by GC/MS selected ion monitoring.

References   -Method under development (PIRA, Leatherhead, UK).
J. Sci. Food Agric., 1991, 54, 549–555.
Analyst, 1988, 13, 335–338.

Acrylamide

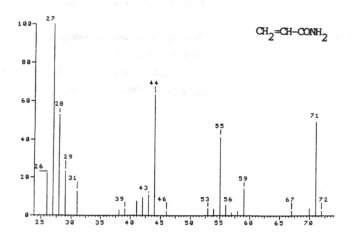

$CH_2=CH-CONH_2$

M/Z	Ion Intensity(%)	M/Z	Ion Intensity(%)
26	22.0	44	63.5
27	100.0	46	2.5
28	53.0	53	3.6
29	23.2	54	3.2
31	12.7	55	41.4
38	2.6	56	5.3
39	3.3	59	14.0
41	7.4	70	3.8
42	9.1	71	49.0
43	10.5	72	2.5

Spectrometer   :Finnigan Mat SSQ 70
Inlet System   :Capillary GC/MS
Source Temperature :150°C
Electron Energy   :70 eV
Scan Range   :25–400

31

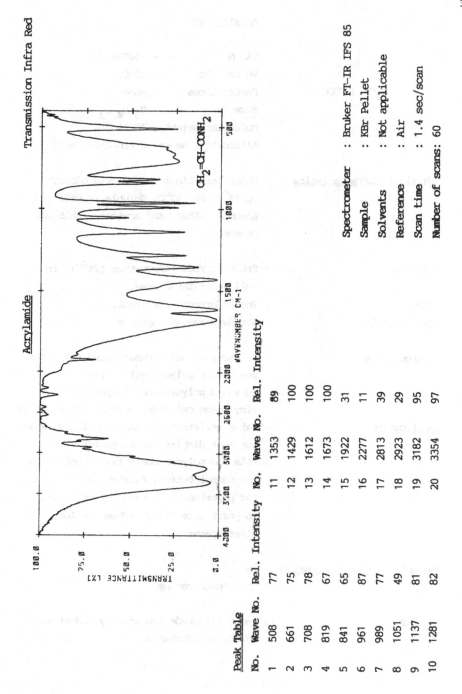

Acrylamide Transmission Infra Red

$CH_2=CH-CONH_2$

Spectrometer	: Bruker FT-IR IFS 85
Sample	: KBr Pellet
Solvents	: Not applicable
Reference	: Air
Scan time	: 1.4 sec/scan
Number of scans: 60	

Peak Table

No.	Wave No.	Rel. Intensity	No.	Wave No.	Rel. Intensity
1	508	77	11	1353	89
2	661	75	12	1429	100
3	708	78	13	1612	100
4	819	67	14	1673	100
5	841	65	15	1922	31
6	961	87	16	2277	11
7	989	77	17	2813	39
8	1051	49	18	2923	29
9	1137	81	19	3182	95
10	1281	82	20	3354	97

Acrylic acid

H₂C=CH–COOH

CAS No.	– 00079–10–7
PM Ref. No.	– 10690
Restrictions	– none
Formula	– $C_3 H_4 O_2$
Molecular weight	– 72.06
Alternative names	– 2–Propenoic acid.

Physical Characteristics
- Colourless liquid with acrid odour, mp 13°C, bp 142°C. Soluble in water, alcohol, ether, and acetone. Inhibitor present.

Handling
- Store at room temperature (25°C), in a well ventillated area.

Safety
- Toxic/Corrosive/Irritant.

Availability
- Standard sample supplied.

Current uses
- Polyacrylic acid. Unsaturated polyester resins. Co-polymer with styrene for expanded polystyrene. Co-polymer with vinylidene chloride. Acrylic latex sprays.

Applications
- Water resistant non-blocking films. Trays for meat display. Coatings for nylon film and polycarbonate film. Containers for peanut butter. Barrier latex coating, for paperboard. Coating for nylon and polycarbonate films. Adhesives for blisterpacks.

Methods of Characterisation
- IR
  Mass Spectroscopy

Purity
- Technical grade supplied by industry. Purity not declared.

33

Acrylic acid

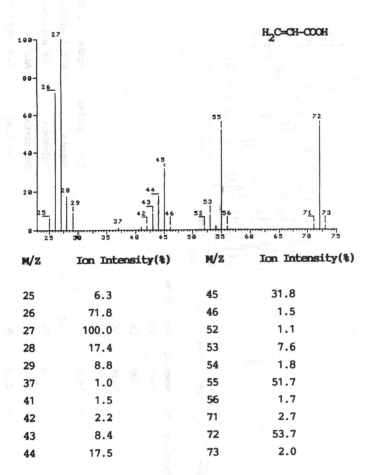

$H_2C=CH-COOH$

M/Z	Ion Intensity(%)	M/Z	Ion Intensity(%)
25	6.3	45	31.8
26	71.8	46	1.5
27	100.0	52	1.1
28	17.4	53	7.6
29	8.8	54	1.8
37	1.0	55	51.7
41	1.5	56	1.7
42	2.2	71	2.7
43	8.4	72	53.7
44	17.5	73	2.0

Spectrometer   :Finnigan Mat SSQ 70
Inlet System   :Capillary GC/MS
Source Temperature:150°C
Electron Energy   :70 eV
Scan Range   :25-400

34

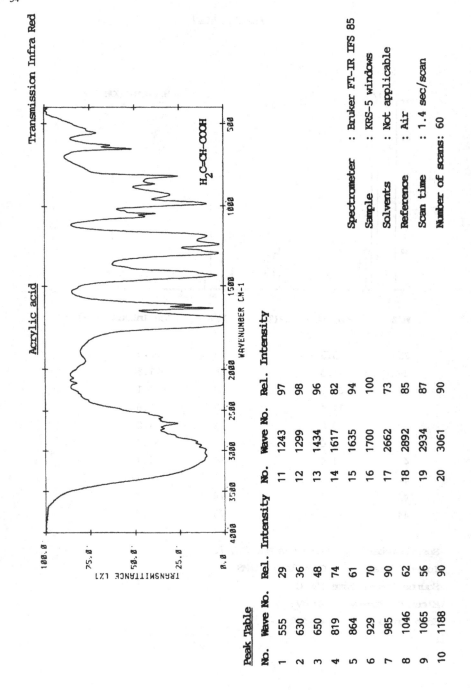

Transmission Infra Red

Acrylic acid

H₂C=CH-COOH

WAVENUMBER CM-1

Spectrometer	: Bruker FT-IR IFS 85
Sample	: KRS-5 windows
Solvents	: Not applicable
Reference	: Air
Scan time	: 1.4 sec/scan
Number of scans: 60	

Peak Table

No.	Wave No.	Rel. Intensity	No.	Wave No.	Rel. Intensity
1	555	29	11	1243	97
2	630	36	12	1299	98
3	650	48	13	1434	96
4	819	74	14	1617	82
5	864	61	15	1635	94
6	929	70	16	1700	100
7	985	90	17	2662	73
8	1046	62	18	2892	85
9	1065	56	19	2934	87
10	1188	90	20	3061	90

Acrylic acid, ethyl ester

$$CH_2=CH-COO-C_2H_5$$

CAS No.   – 00140-88-5
PM Ref. No.   – 11470
Restrictions   – none
Formula   – $C_5 H_8 O_2$
Molecular weight – 100.12
Alternative names– Ethyl acrylate;
2-propenoic acid, ethyl
ester.

Physical Characteristics   – Colourless, pungent liquid, mp -72°C,
bp 100°C. Soluble in alcohol, ether, and
chloroform. Inhibited with 15 mg/kg
hydroquinone monomethyl ether.

Handling   – Store at room temperature (25°C).
Safety   – Flammable/Irritant.
Availability   – Standard sample supplied.

Current uses   – Polyethyl acrylate. Co-polymer with
ethylene (EEA) and vinylidene chloride.
Cross-linking agent for polyester resins.
Used in elastomeric latexes.

Applications   – Coating for nylon film. Rigid and
semi-rigid repeat use containers. Resin
bonded filters for milk, potable water
etc. Adhesive for blisterpacks.

Methods of Characterisation – IR
Mass Spectroscopy

Purity   – 99.5%

Acrylic acid, ethyl ester

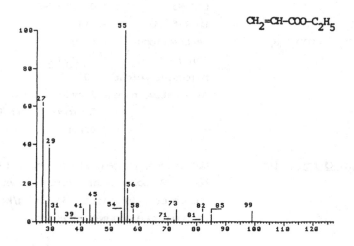

$CH_2=CH-COO-C_2H_5$

M/Z	Ion Intensity(%)	M/Z	Ion Intensity(%)
27	59.0	53	2.3
28	10.5	54	5.7
29	38.2	55	100.0
30	2.4	56	13.8
31	2.5	57	1.2
41	2.3	58	3.7
42	1.8	73	6.5
43	9.0	82	4.0
44	2.3	85	3.6
45	10.3	99	5.5

Spectrometer   :Finnigan Mat SSQ 70
Inlet System   :Capillary GC/MS
Source Temperature:150°C
Electron Energy   :70 eV
Scan Range   :25–400

Acrylic acid, ethyl ester   Transmission Infra Red

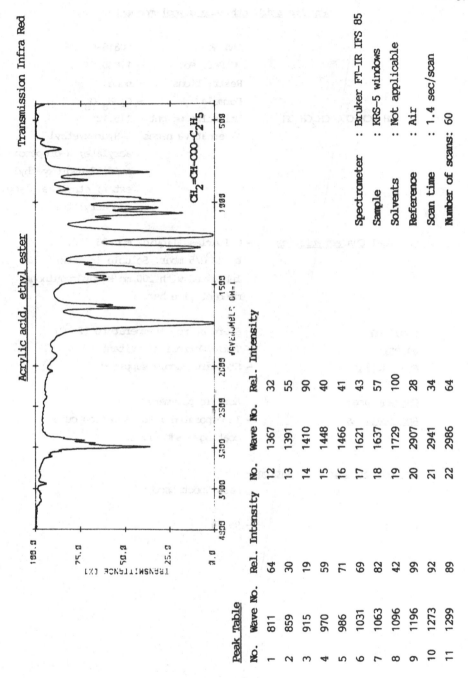

$CH_2=CH-COO-C_2H_5$

Spectrometer	: Bruker FT-IR IFS 85
Sample	: KRS-5 windows
Solvents	: Not applicable
Reference	: Air
Scan time	: 1.4 sec/scan
Number of scans: 60	

Peak Table

No.	Wave No.	Rel. Intensity	No.	Wave No.	Rel. Intensity
1	811	64	12	1367	32
2	859	30	13	1391	55
3	915	19	14	1410	90
4	970	59	15	1448	40
5	986	71	16	1466	41
6	1031	69	17	1621	43
7	1063	82	18	1637	57
8	1096	42	19	1729	100
9	1196	99	20	2907	28
10	1273	92	21	2941	34
11	1299	89	22	2986	64

Acrylic acid, ethylene glycol monoester

$$CH_2=CH-COO-CH_2CH_2OH$$

CAS No.	– 00818–61–1
PM Ref. No.	– 11830
Restrictions	– none
Formula	– $C_5 H_8 O_3$
Molecular weight	– 116.12
Alternative names	– 2–Hydroxyethyl acrylate; 2–propenoic acid, 2–hydroxyethyl ester; ethylene glycol monoacrylate.

Physical Characteristics   – Colourless liquid, mp –35.5^{O}C, bp 82°C/5 mbar. Soluble in water. Inhibited with 200 mg/kg hydroquinone monomethyl ether.

Handling   – Store at room temperature (25°C).

Safety   – Toxic/Corrosive/Irritant.

Availability   – Standard sample supplied.

Current uses   – Acrylic polymers.

Applications   – Incorporated into radiation curable coatings. Adhesives.

Methods of Characterisation – IR
Mass Spectroscopy

Purity   – 95.7%

39

Acrylic acid, ethyleneglycol monoester

$$CH_2=CH-COO-CH_2CH_2OH$$

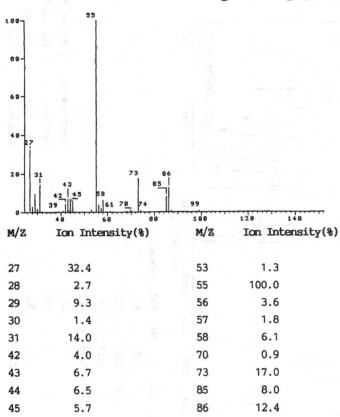

M/Z	Ion Intensity(%)	M/Z	Ion Intensity(%)
27	32.4	53	1.3
28	2.7	55	100.0
29	9.3	56	3.6
30	1.4	57	1.8
31	14.0	58	6.1
42	4.0	70	0.9
43	6.7	73	17.0
44	6.5	85	8.0
45	5.7	86	12.4

Spectrometer   :Finnigan Mat SSQ 70
Inlet System   :Capillary GC/MS
Source Temperature:150°C
Electron Energy   :70 eV
Scan Range   :25-400

40

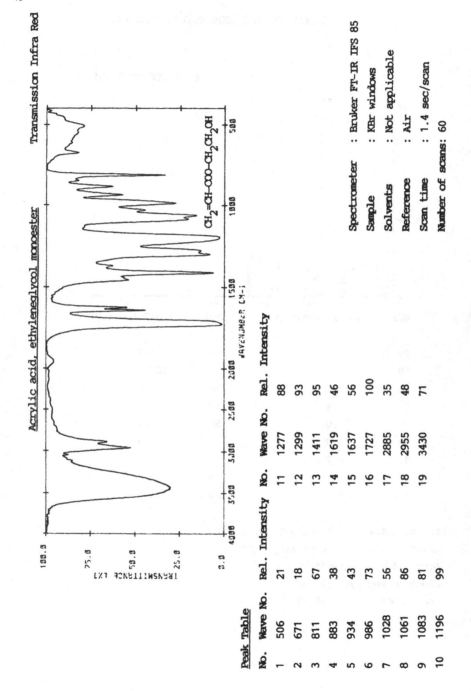

Acrylic acid, ethyleneglycol monoester   Transmission Infra Red

CH$_2$=CH-OOO-CH$_2$CH$_2$OH

Spectrometer : Bruker FT-IR IFS 85
Sample   : KBr windows
Solvents   : Not applicable
Reference   : Air
Scan time   : 1.4 sec/scan
Number of scans: 60

Peak Table

No.	Wave No.	Rel. Intensity	No.	Wave No.	Rel. Intensity
1	506	21	11	1277	88
2	671	18	12	1299	93
3	811	67	13	1411	95
4	883	38	14	1619	46
5	934	43	15	1637	56
6	986	73	16	1727	100
7	1028	56	17	2885	35
8	1061	86	18	2955	48
9	1083	81	19	3430	71
10	1196	99			

Acrylic acid, isobutyl ester

$$CH_2=CH-COO-CH_2-CH(CH_3)_2$$

CAS No.	– 00106–63–8
PM Ref. No.	– 11590
Restrictions	– none
Formula	– $C_7 H_{12} O_2$
Molecular weight	– 128.17
Alternative names	– Isobutyl acrylate.

Physical Characteristics — Liquid, bp 139°C. Inhibited with 50 mg/kg hydroquinone monomethyl ether.

Handling — Store at room temperature (25°C).
Safety — Harmful.
Availability — Standard sample supplied.

Current uses — Used in acrylate polymers. In the synthesis of polyvinylidene chloride, polyvinyl chloride, polyethylene and polystyrene. Ionomeric resins with methacrylic acid and ethylene.

Applications — Used in films, coatings and adhesives. Rigid and semi-rigid containers.

Methods of Characterisation — IR
Mass Spectroscopy

Purity — >99%

Acrylic acid, isobutyl ester

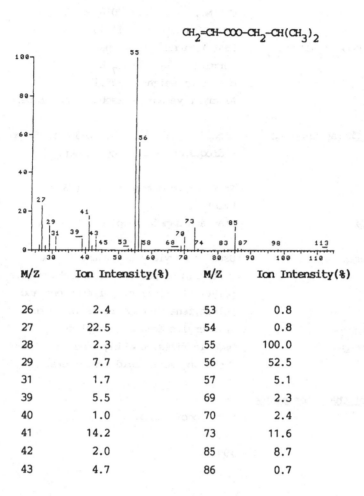

$$CH_2=CH-COO-CH_2-CH(CH_3)_2$$

M/Z	Ion Intensity(%)	M/Z	Ion Intensity(%)
26	2.4	53	0.8
27	22.5	54	0.8
28	2.3	55	100.0
29	7.7	56	52.5
31	1.7	57	5.1
39	5.5	69	2.3
40	1.0	70	2.4
41	14.2	73	11.6
42	2.0	85	8.7
43	4.7	86	0.7

Spectrometer   :Finnigan Mat SSQ 70
Inlet System   :Capillary GC/MS
Source Temperature:150°C
Electron Energy   :70 eV
Scan Range   :25–400

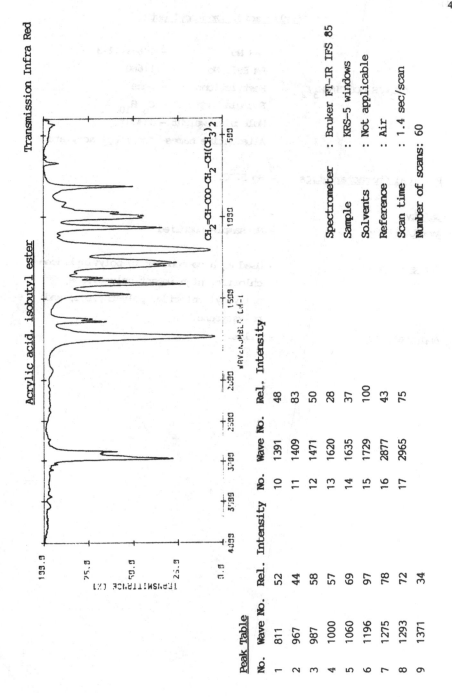

Acrylic acid, isobutyl ester

Transmission Infra Red

$CH_2=CH-COO-CH_2-CH(CH_3)_2$

Spectrometer	: Bruker FT-IR IFS 85
Sample	: KRS-5 windows
Solvents	: Not applicable
Reference	: Air
Scan time	: 1.4 sec/scan
Number of scans: 60	

Peak Table

No.	Wave No.	Rel. Intensity	No.	Wave No.	Rel. Intensity
1	811	52	10	1391	48
2	967	44	11	1409	83
3	987	58	12	1471	50
4	1000	57	13	1620	28
5	1060	69	14	1635	37
6	1196	97	15	1729	100
7	1275	78	16	2877	43
8	1293	72	17	2965	75
9	1371	34			

Acrylic acid, isopropyl ester

$$CH_2=CH-COOCH(CH_3)_2$$

CAS No.	– 00689-12-3
PM Ref. No.	– 11680
Restrictions	– none
Formula	– $C_6 H_{10} O_2$
Molecular weight	– 114.15
Alternative names	– Isopropyl acrylate.

Physical Characteristics — Bp 52°C.

Safety —

Availability — No sample supplied.

Current uses — Used as a co-monomer in polyvinylidene chloride, high impact polystyrene, polyvinyl chloride, polyethylene, and polystyrene.

Applications — Films.

Acrylic acid, methyl ester

$$CH_2=CH-COOCH_3$$

CAS No.	– 00096–33–3
PM Ref. No.	– 11710
Restrictions	– none
Formula	– $C_4 H_6 O_2$
Molecular weight	– 86.09
Alternative names	– Methyl acrylate; 2–propenoic acid, methyl ester.

Physical Characteristics — Colourless pungent liquid, mp –75°C, bp 80°C. Soluble in alcohol and ether. Inhibited with 15–20 mg/kg hydroquinone monomethyl ether.

Handling — Refrigerate (4°C).
Safety — Flammable/Irritant.
Availability — Standard sample supplied.

Current uses — Co–polymer with vinylidene chloride. Co–polymer resins with ethylene. Cross–linking agent for polyester resins. Used in elastomeric latexes.

Applications — Coating for nylon film and polycarbonate film. Paper coatings.

Methods of Characterisation — IR
Mass Spectroscopy

Purity — 99.5%

Acrylic acid, methyl ester

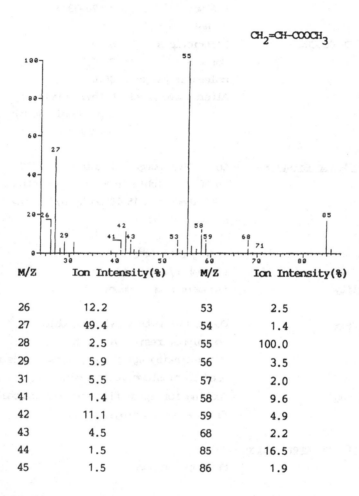

$CH_2=CH-COOCH_3$

M/Z	Ion Intensity(%)	M/Z	Ion Intensity(%)
26	12.2	53	2.5
27	49.4	54	1.4
28	2.5	55	100.0
29	5.9	56	3.5
31	5.5	57	2.0
41	1.4	58	9.6
42	11.1	59	4.9
43	4.5	68	2.2
44	1.5	85	16.5
45	1.5	86	1.9

Spectrometer   :Finnigan Mat SSQ 70
Inlet System   :Capillary GC/MS
Source Temperature:150°C
Electron Energy   :70 eV
Scan Range   :25-400

47

Acrylic acid, methyl ester   Transmission Infra Red

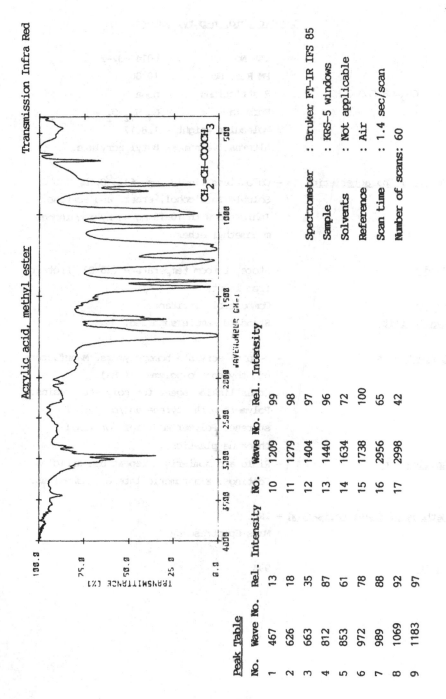

CH$_2$=CH-COOCH$_3$

Spectrometer : Bruker FT-IR IFS 85

Sample   : KRS-5 windows

Solvents   : Not applicable

Reference   : Air

Scan time   : 1.4 sec/scan

Number of scans: 60

Peak Table

No.	Wave No.	Rel. Intensity	No.	Wave No.	Rel. Intensity
1	467	13	10	1209	99
2	626	18	11	1279	98
3	663	35	12	1404	97
4	812	87	13	1440	96
5	853	61	14	1634	72
6	972	78	15	1738	100
7	989	88	16	2956	65
8	1069	92	17	2998	42
9	1183	97			

Acrylic acid, n-butyl ester

$CH_2=CH-COO-C_4H_9$

CAS No.   – 00141–32–2
PM Ref. No.   – 10780
Restrictions   – none
Formula   – $C_7 H_{12} O_2$
Molecular weight – 128.17
Alternative names– Butyl acrylate.

Physical Characteristics   – Colourless liquid, mp –64.6^{O}C, bp 145°C.
Soluble in alcohol, ether, and acetone.
Inhibited with 10–15 mg/kg hydroquinone
monomethyl ether.

Handling   – Store at room temperature (25°C). Protect
from light.

Safety   – Combustible/Irritant.

Availability   – Standard sample supplied.

Current uses   – n–Butyl acrylate homopolymers. Manufacture
of ethylene co–polymers (EBA).
Cross–linking agent for polyester resins.
Polymers with styrene and/or methyl
styrene. Polymer modifier for vinyl
chloride plastics.

Applications   – Rigid and semi–rigid repeat use articles.
Coatings. Elastomeric latexes. Adhesives.

Methods of Characterisation – IR
Mass Spectroscopy

Purity   – 99%

Acrylic acid, n-butyl ester

$CH_2=CH-COO-C_4H_9$

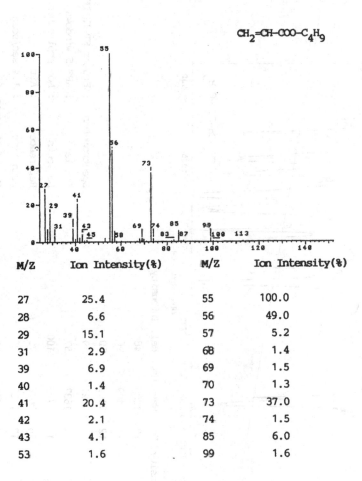

M/Z	Ion Intensity(%)	M/Z	Ion Intensity(%)
27	25.4	55	100.0
28	6.6	56	49.0
29	15.1	57	5.2
31	2.9	68	1.4
39	6.9	69	1.5
40	1.4	70	1.3
41	20.4	73	37.0
42	2.1	74	1.5
43	4.1	85	6.0
53	1.6	99	1.6

Spectrometer   :Finnigan Mat SSQ 70
Inlet System   :Capillary GC/MS
Source Temperature:150°C
Electron Energy   :70 eV
Scan Range   :25-400

50

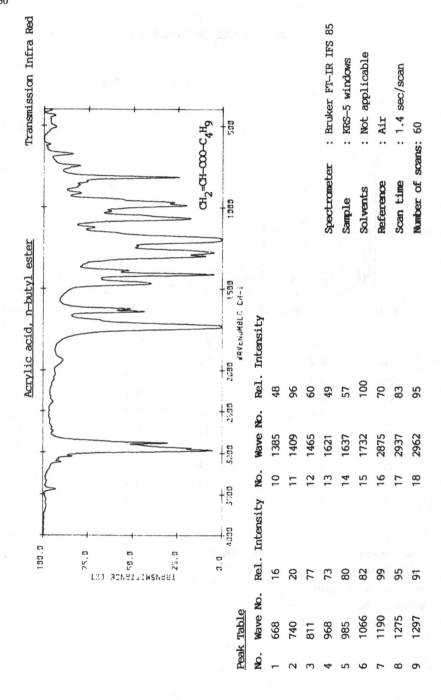

Acrylic acid, n-butyl ester

Transmission Infra Red

CH_2=CH-OOO-C_4H_9

Spectrometer : Bruker FT-IR IFS 85
Sample : KRS-5 windows
Solvents : Not applicable
Reference : Air
Scan time : 1.4 sec/scan
Number of scans: 60

Peak Table

No.	Wave No.	Rel. Intensity	No.	Wave No.	Rel. Intensity
1	668	16	10	1385	48
2	740	20	11	1409	96
3	811	77	12	1465	60
4	968	73	13	1621	49
5	985	80	14	1637	57
6	1066	82	15	1732	100
7	1190	99	16	2875	70
8	1275	95	17	2937	83
9	1297	91	18	2962	95

Acrylic acid, propyl ester

$CH_2=CH-COOCH_2CH_2CH_3$

CAS No.	- 00925-60-0
PM Ref. No.	- 11980
Restrictions	- none
Formula	- $C_6 H_{10} O_2$
Molecular weight	- 114.15
Alternative names	-Propyl acrylate.

Physical Characteristics   - Bp 44°C.

Safety   -

Availability   - No sample supplied.

Current uses   - Used as a co-monomer in polyvinylidene
chloride, high impact polystyrene,
polyvinyl chloride, polyethylene, and
polystyrene.

Applications   - Films.

Acrylic acid, sec butyl ester

$CH_2=CH-COOCH(CH_3)CH_2CH_3$

CAS No.   - 02998-08-5
PM Ref. No.   - 10810
Restrictions   - none
Formula   - $C_7 H_{12} O_2$
Molecular weight - 128.17
Alternative names-

Physical Characteristics   - Bp 60°C.

Safety   -
Availability   - No sample supplied.

Current uses   - Used as a co-monomer in polyvinylidene chloride, high impact polystyrene, polyvinyl chloride, polyethylene, and polystyrene.

Applications   - Films.

Acrylic acid, tert. butyl ester

$CH_2=CH-COO-C(CH_3)_3$

CAS No.	- 01663-39-4
PM Ref. No.	- 10840
Restrictions	- none
Formula	- $C_7 H_{12} O_2$
Molecular weight	- 128.17
Alternative names	- tert-butyl acrylate.

Physical Characteristics - Colourless liquid, bp 61-63°C/0.08 bar. Stabilised with 50-75 mg/kg hydroquinone monomethyl ether.

Handling - Store at room temperature (25°C).

Safety - Flammable/Irritant.

Availability - Standard sample supplied.

Current uses - Used in rigid and semi-rigid acrylic plastics.

Applications - Repeat-use containers.

Methods of Characterisation - IR
Mass Spectroscopy

Purity - 98%

Acrylic acid, tert-butyl ester

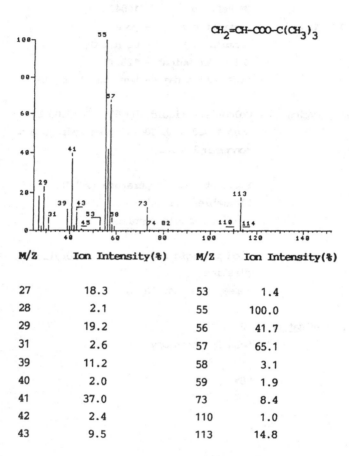

$CH_2=CH-COO-C(CH_3)_3$

M/Z	Ion Intensity(%)	M/Z	Ion Intensity(%)
27	18.3	53	1.4
28	2.1	55	100.0
29	19.2	56	41.7
31	2.6	57	65.1
39	11.2	58	3.1
40	2.0	59	1.9
41	37.0	73	8.4
42	2.4	110	1.0
43	9.5	113	14.8

Spectrometer   :Finnigan Mat SSQ 70
Inlet System   :Capillary GC/MS
Source Temperature:150°C
Electron Energy   :70 eV
Scan Range   :25-400

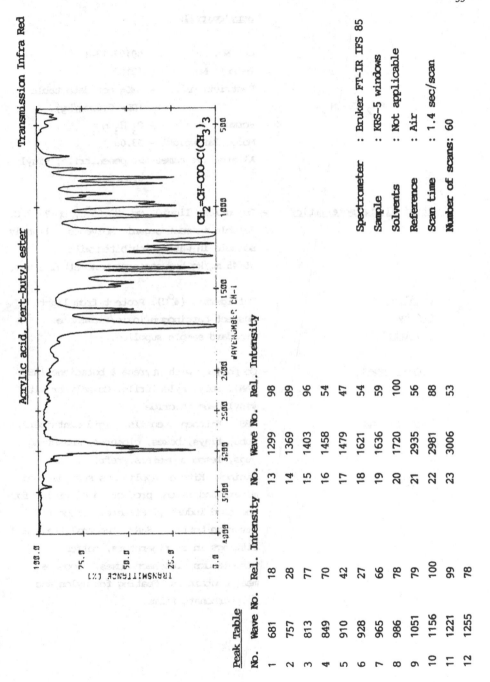

Acrylic acid, tert-butyl ester   Transmission Infra Red

$CH_2=CH-COO-C(CH_3)_3$

Spectrometer	: Bruker FT-IR IFS 85
Sample	: KRS-5 windows
Solvents	: Not applicable
Reference	: Air
Scan time	: 1.4 sec/scan
Number of scans: 60	

Peak Table

No.	Wave No.	Rel. Intensity	No.	Wave No.	Rel. Intensity
1	681	18	13	1299	98
2	757	28	14	1369	89
3	813	77	15	1403	96
4	849	70	16	1458	54
5	910	42	17	1479	47
6	928	27	18	1621	54
7	965	66	19	1636	59
8	986	78	20	1720	100
9	1051	79	21	2935	56
10	1156	100	22	2981	88
11	1221	99	23	3006	53
12	1255	78			

Acrylonitrile

CH₂=CH-CN

CAS No.	– 00107-13-1
PM Ref. No.	– 12100
Restrictions	– SML= not detectable (DL= 0.02mg/kg).
Formula	– $C_3 H_3 N$
Molecular weight	– 53.06
Alternative names	– Propenenitrile; vinyl cyanide.

Physical Characteristics
– Colourless liquid, mp $-83.55^{\circ}C$, bp $77.3^{\circ}C$. Soluble in most organic solvents, slightly soluble in water. Inhibited with 35-45 mg/kg hydroquinone monomethyl ether.

Handling
– Refrigerate ($4^{\circ}C$). Protect from light.

Safety
– Suspect carcinogen/Toxic/Flammable.

Availability
– Standard sample supplied.

Current uses
– Co-polymer with styrene & butadiene (ABS, SAN). Polyacrylonitrile. Co-polymer with vinylidene chloride.

Applications
– ABS – Kitchen utensils, rigid containers, tubs, trays, boxes, closures, measuring jugs, lemon squeezers, refrigerator linings. Kitchen appliances such as food mixers and in the production of piping for the food industry. Filters. Carbonated beverage bottles. SAN – Internal trays and fittings in refrigerators, coffee percolators, luncheon boxes. Processed meat containers. Coating for nylon and polycarbonate films.

Methods of Characterisation - IR
                              Mass Spectroscopy

Purity   - 99%

Analytical methods   - Headspace GC with nitrogen-selective
                                detection. Foods or simulants equilibrated
                                at 70°C prior to headspace sampling.
                                Proprionitrile used as internal standard.
                                Confirmation by GC/MS selected ion
                                monitoring.

References   - Draft CEN method (Fraunhofer Inst. Munich,
                                D).
                                Food Chem., 1982, 9, 234-252.
                                J. Assoc. Off. Anal. Chem., 1985, 68, 776.

Acrylonitrile

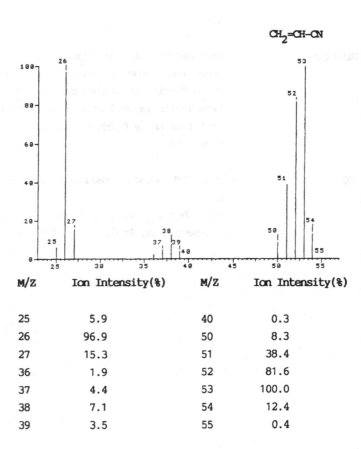

$CH_2=CH-CN$

M/Z	Ion Intensity(%)	M/Z	Ion Intensity(%)
25	5.9	40	0.3
26	96.9	50	8.3
27	15.3	51	38.4
36	1.9	52	81.6
37	4.4	53	100.0
38	7.1	54	12.4
39	3.5	55	0.4

Spectrometer   :Finnigan Mat SSQ 70
Inlet System   :Capillary GC/MS
Source Temperature:150°C
Electron Energy   :70 eV
Scan Range   :25–400

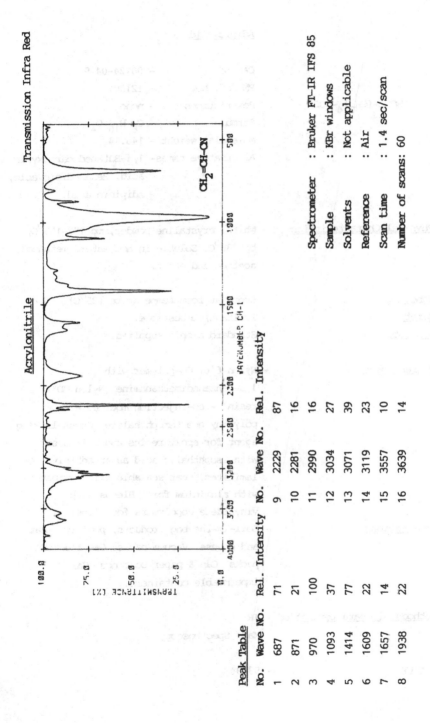

Acrylonitrile

Transmission Infra Red

$CH_2=CH-CN$

Spectrometer	: Bruker FT-IR IFS 85
Sample	: KBr windows
Solvents	: Not applicable
Reference	: Air
Scan time	: 1.4 sec/scan
Number of scans: 60	

Peak Table

No.	Wave No.	Rel. Intensity	No.	Wave No.	Rel. Intensity
1	687	71	9	2229	87
2	871	21	10	2281	16
3	970	100	11	2990	16
4	1093	37	12	3034	27
5	1414	77	13	3071	39
6	1609	22	14	3119	23
7	1657	14	15	3557	10
8	1938	22	16	3639	14

Adipic acid

HOOC-(CH$_2$)$_4$-COOH

CAS No.	– 00124-04-9
PM Ref. No.	– 12130
Restrictions	– none
Formula	– C$_6$ H$_{10}$ O$_4$
Molecular weight	– 146.14
Alternative names	– 1,4-Butanedicarboxylic acid, Hexanedioic acid, Adipinic acid.

Physical Characteristics
— White, crystaline powder, mp 152-153°C, bp 336°C. Soluble in hot water, methanol, acetone and ether.

Handling
— Store at room temperature (25°C).

Safety
— Irritant/Combustible.

Availability
— Standard sample supplied.

Current uses
— Nylon 6'6. Co-polymer with 1,3-benzenedimethanamine (Nylon MDX-6 resin) - co-injection moulded with polyethylene terephthalate. Cross-linking agent for epoxy resins and polyesters. With Isophthalic acid as an adhesive for laminates. Heat sealable film laminates with aluminium foil. Blends with vinylidene copolymers for films.

Applications
— Boil-in-the-bag products, poultry, meat and cheese. Vacuum packs. Gas flushed packs. Can & paper coatings. Gas impermeable containers.

Methods of Characterisation
— IR
Mass Spectroscopy

Purity
— 99.9%

Adipic acid

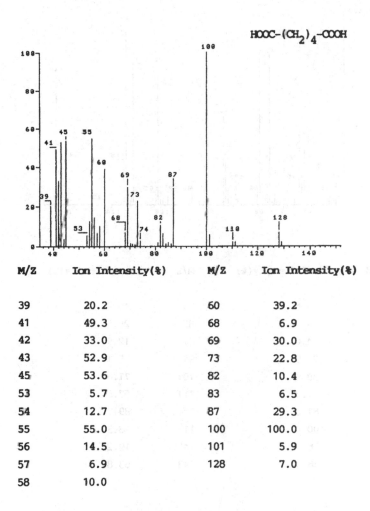

HOOC-(CH$_2$)$_4$-COOH

M/Z	Ion Intensity(%)	M/Z	Ion Intensity(%)
39	20.2	60	39.2
41	49.3	68	6.9
42	33.0	69	30.0
43	52.9	73	22.8
45	53.6	82	10.4
53	5.7	83	6.5
54	12.7	87	29.3
55	55.0	100	100.0
56	14.5	101	5.9
57	6.9	128	7.0
58	10.0		

Spectrometer   :Finnigan Mat SSQ 70
Inlet System   :Capillary GC/MS
Source Temperature:150°C
Electron Energy   :70 eV
Scan Range   :25-400

Adipic acid, dimethyl ester

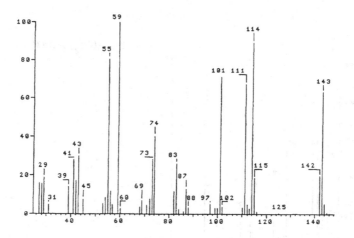

M/Z	Ion Intensity(%)	M/Z	Ion Intensity(%)
29	18.8	74	40.4
31	4.9	83	26.0
39	14.0	87	12.9
41	27.9	97	5.1
43	30.3	101	71.8
45	7.8	111	67.8
55	81.0	114	89.7
59	100.0	115	18.3
69	6.9	142	19.2
73	28.3	143	63.8

Spectrometer   :Finnigan Mat SSQ 70
Inlet System   :Capillary GC/MS
Source Temperature:150°C
Electron Energy   :70 eV
Scan Range   :25-400

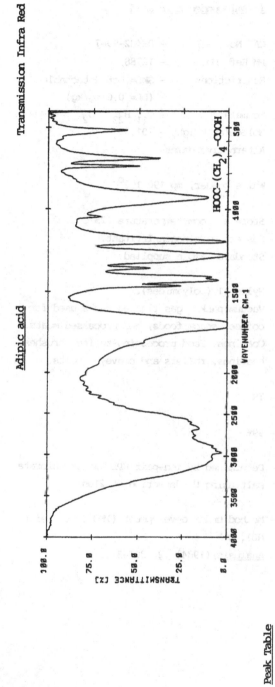

Adipic acid   Transmission Infra Red

HOOC—(CH$_2$)$_4$—COOH

Spectrometer : Bruker FT-IR IFS 85
Sample   : KBr Pellet
Solvents   : Not applicable
Reference   : Air
Scan time   : 1.4 sec/scan
Number of scans: 60

Peak Table

No.	Wave No.	Rel. Intensity	No.	Wave No.	Rel. Intensity
1	515	45	9	1428	95
2	690	63	10	1463	79
3	735	85	11	1700	100
4	1044	31	12	2596	80
5	1194	98	13	2671	85
6	1281	99	14	2754	83
7	1356	59	15	2962	97
8	1409	93	16	3040	92

11-Aminoundecanoic acid

$$H_2N-(CH_2)_{10}-COOH$$

CAS No.	– 02432–99–7
PM Ref. No.	– 12788
Restrictions	– SML= not detectable
	(DL= 0.01mg/kg)
Formula	– $C_{11} H_{23} N O_2$
Molecular weight	– 201.31
Alternative names–	

Physical Characteristics   – White powder, mp 190–192°C.

Handling   – Store at room temperature (25°C).

Safety   – Suspect carcinogen/Irritant.

Availability   – Standard sample supplied.

Current uses   – Nylon 11 (polyamides).

Applications   – Vacuum packs, gas flushed packs used for coffee, snack foods, and processed meats. Coatings. Food processing–tubing, brushes, bearings, rollers and conveyer belts.

Methods of Characterisation – IR

Purity   – 99%

Analytical methods   – Determined by ion–pair HPLC as perchlorate salt using UV detection at 210nm.

References   – Method under development (TNO Inst. Zeist, NL).

Analusis (1984) 12, 307–311.

11-Aminoundecanoic acid   Transmission Infra Red

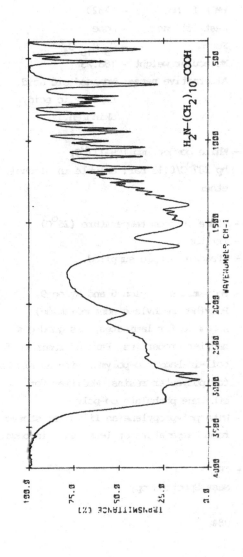

H$_2$N–(CH$_2$)$_{10}$–COOH

Spectrometer : Bruker FT-IR IFS 85
Sample   : KBr Pellet
Solvents   : Not applicable
Reference   : Air
Scan time   : 1.4 sec/scan
Number of scans: 60

Peak Table

No.	Wave No.	Rel. Intensity	No.	Wave No.	Rel. Intensity
1	445	65	11	1039	68
2	516	44	12	1392	94
3	657	75	13	1458	87
4	724	52	14	1499	90
5	750	48	15	1642	88
6	786	37	16	2115	57
7	830	38	17	2540	69
8	889	37	18	2851	98
9	944	60	19	2924	100
10	999	46	20	3155	80

Azelaic acid

HOOC-(CH$_2$)$_7$-COOH

CAS No.	– 00123-99-9
PM Ref. No.	– 12820
Restrictions	– none
Formula	– C$_9$ H$_{16}$ O$_4$
Molecular weight – 188.23	
Alternative names– Nonanedioic acid,
	Lepargylic acid,
	Skinoren.

Physical Characteristics   – White powder, mp 109–111°C, bp 287°C/0.13 bar. Soluble in alcohol, and ether.

Handling   – Store at room temperature (25°C).

Safety   – Irritant.

Availability   – Standard sample supplied.

Current uses   – Polyamides - Nylon 6 and Nylon 9, Poly(hexamethylene nonanediamide). Adhesive for laminates, has good gas barrier properties. Part of laminate for boil-in-bags. Co-polymer with an alcohol for polyester resins. Modifier for ethylene phthalate co-polymers.

Applications   – With polypropylene as films for storage at room temperature or less. Boil-in-bags.

Methods of Characterisation   – IR

Mass Spectroscopy

Purity   – 98%

Azelaic acid

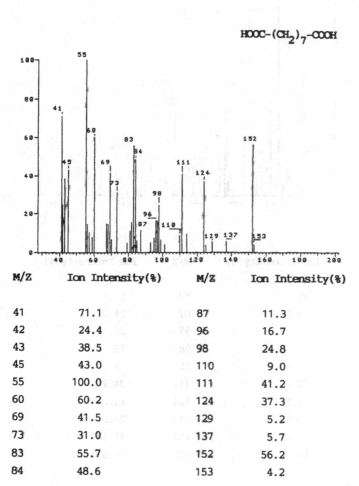

$HOOC-(CH_2)_7-COOH$

M/Z	Ion Intensity(%)	M/Z	Ion Intensity(%)
41	71.1	87	11.3
42	24.4	96	16.7
43	38.5	98	24.8
45	43.0	110	9.0
55	100.0	111	41.2
60	60.2	124	37.3
69	41.5	129	5.2
73	31.0	137	5.7
83	55.7	152	56.2
84	48.6	153	4.2

Spectrometer   :Finnigan Mat SSQ 70
Inlet System   :Capillary GC/MS
Source Temperature:150°C
Electron Energy   :70 eV
Scan Range   :25–400

Azelaic acid, dimethyl ester

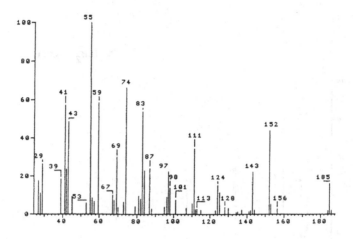

M/Z	Ion Intensity(%)	M/Z	Ion Intensity(%)
29	26.4	83	54.1
39	18.5	87	24.2
41	57.3	97	22.5
43	48.6	98	13.8
53	6.0	101	7.5
55	100.0	111	34.7
59	58.8	124	15.2
67	9.9	143	22.3
69	30.4	152	44.3
74	66.3	185	16.3

Spectrometer   :Finnigan Mat SSQ 70
Inlet System   :Capillary GC/MS
Source Temperature :150°C
Electron Energy   :70 eV
Scan Range   :25–400

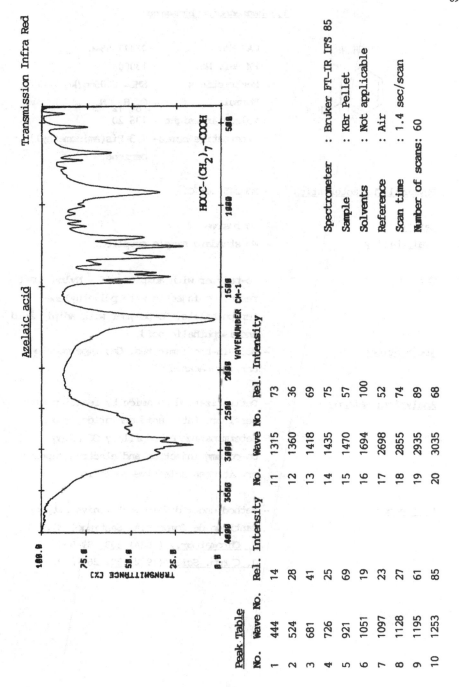

Azelaic acid   Transmission Infra Red

HOOC-(CH$_2$)$_7$-COOH

Spectrometer : Bruker FT-IR IFS 85
Sample   : KBr Pellet
Solvents   : Not applicable
Reference   : Air
Scan time   : 1.4 sec/scan
Number of scans: 60

Peak Table

No.	Wave No.	Rel. Intensity	No.	Wave No.	Rel. Intensity
1	444	14	11	1315	73
2	524	28	12	1360	36
3	681	41	13	1418	69
4	726	25	14	1435	75
5	921	69	15	1470	57
6	1051	19	16	1694	100
7	1097	23	17	2698	52
8	1128	27	18	2855	74
9	1195	61	19	2935	89
10	1253	85	20	3035	68

1,3-Benzenedimethanamine

CH$_2$NH$_2$

CH$_2$NH$_2$

CAS No.	– 01477-55-0
PM Ref. No.	– 13000
Restrictions	– SML= 0.05mg/kg
Formula	– C$_8$ H$_{12}$ N$_2$
Molecular weight	– 136.20
Alternative names	– 1,3-Bis(aminomethyl) benzene.

Physical Characteristics   – Bp 245-248°C.

Safety   – Corrosive.

Availability   – No standard sample supplied.

Current uses   – Co-polymer with adipic acid – Nylon MDX-6 resin. Co-injected with polyethylene terephthalate. Co-polymer with adipic acid and isophthalic acid.

Applications   – Boil-in-bag laminates. Gas impermeable barrier packs.

Analytical methods   – Derivatization to amide by reaction with perfluoro fatty acid anhydride, and determination by capillary GC using on-column injection and electron capture or nitrogen selective-detection.

References   – Method under development (University of Santiago de Composela, Santiago, E). J. Chromatogr., (1984) 303, 89-98. J. Chrom. Sci., (1992) 30, 267-270.

Benzoic acid

COOH

CAS No.	— 00065–85–0
PM Ref. No.	— 13090
Restrictions	— none
Formula	— $C_7 H_6 O_2$
Molecular weight	— 122.12
Alternative names	— Benzenecarboxylic acid, Phenyl formic acid.

Physical Characteristics	— White crystalline powder, mp 122–123°C, bp 250°C. Soluble in aqeous alkali, ether, acetone, and benzene.
Handling	— Store at room temperature (25°C).
Safety	— Irritant.
Availability	— Standard sample supplied.
Current uses	— Acrylic co-polymer. Unsaturated polyester resins. Epoxy resins. In polyamides. In the manufacture of vulcanised in rubber.
Applications	— Coating aluminium. Can coatings (beer). Milk tubing.
Methods of Characterisation	— IR Mass Spectroscopy
Purity	— 99.5%

Benzoic acid

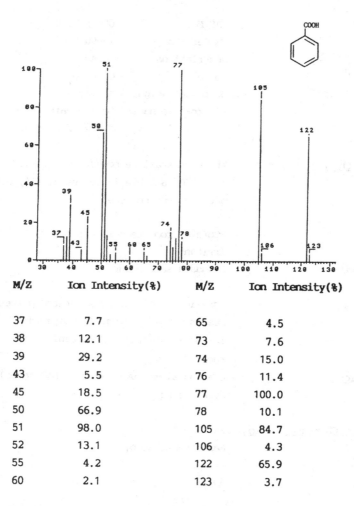

M/Z	Ion Intensity(%)	M/Z	Ion Intensity(%)
37	7.7	65	4.5
38	12.1	73	7.6
39	29.2	74	15.0
43	5.5	76	11.4
45	18.5	77	100.0
50	66.9	78	10.1
51	98.0	105	84.7
52	13.1	106	4.3
55	4.2	122	65.9
60	2.1	123	3.7

Spectrometer   :Finnigan Mat SSQ 70
Inlet System   :Capillary GC/MS
Source Temperature:150°C
Electron Energy   :70 eV
Scan Range   :25–400

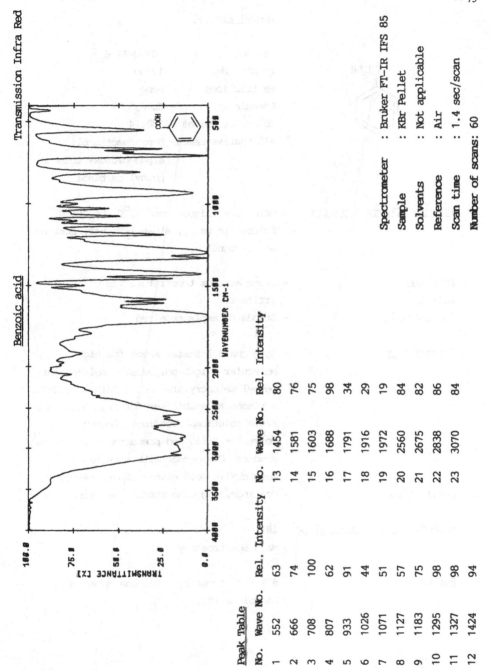

Benzoic acid

Transmission Infra Red

Spectrometer	: Bruker FT-IR IFS 85
Sample	: KBr Pellet
Solvents	: Not applicable
Reference	: Air
Scan time	: 1.4 sec/scan
Number of scans	: 60

Peak Table

No.	Wave No.	Rel. Intensity	No.	Wave No.	Rel. Intensity
1	552	63	13	1454	80
2	666	74	14	1581	76
3	708	100	15	1603	75
4	807	62	16	1688	98
5	933	91	17	1791	34
6	1026	44	18	1916	29
7	1071	51	19	1972	19
8	1127	57	20	2560	84
9	1183	75	21	2675	82
10	1295	98	22	2838	86
11	1327	98	23	3070	84
12	1424	94			

Benzyl alcohol

CH$_2$OH

CAS No.	— 00100–51–6
PM Ref. No.	— 13150
Restrictions	— none
Formula	— C$_7$ H$_8$ O
Molecular weight	— 108.14
Alternative names	— Benzenemethanol, alpha–hydroxy toluene, phenyl carbinol.

Physical Characteristics — Colourless liquid, mp –15°C, bp 205°C. Soluble in water, ethanol, ether, acetone and benzene.

Handling — Store at room temperature (25°C).
Safety — Irritant.
Availability — Standard sample supplied.

Current uses — As a chain transfer agent for high molecular weight polymers. A solvent for methyl methacrylate and cellulose acetate. A viscosity stabiliser for organosilicone resin solutions. Bonding oriented polyester films and preparing nylon resin powders. Co–monomer with acrylic and methacrylic acid esters. Epoxy resins.

Applications — Coatings, e.g wine storage vessels.

Methods of Characterisation — IR
Mass Spectroscopy

Purity — 83% (14% impurity – toluene ester of benzoic acid).

Benzyl alcohol

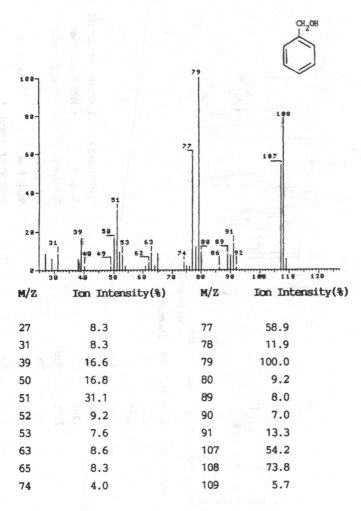

M/Z	Ion Intensity(%)	M/Z	Ion Intensity(%)
27	8.3	77	58.9
31	8.3	78	11.9
39	16.6	79	100.0
50	16.8	80	9.2
51	31.1	89	8.0
52	9.2	90	7.0
53	7.6	91	13.3
63	8.6	107	54.2
65	8.3	108	73.8
74	4.0	109	5.7

Spectrometer   :Finnigan Mat SSQ 70
Inlet System   :Capillary GC/MS
Source Temperature:150°C
Electron Energy   :70 eV
Scan Range   :25–400

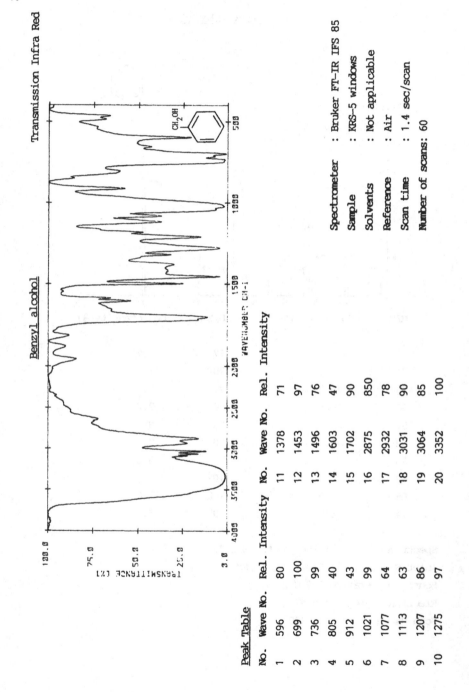

Benzyl alcohol

Transmission Infra Red

Spectrometer	: Bruker FT-IR IFS 85
Sample	: KRS-5 windows
Solvents	: Not applicable
Reference	: Air
Scan time	: 1.4 sec/scan
Number of scans: 60	

Peak Table

No.	Wave No.	Rel. Intensity	No.	Wave No.	Rel. Intensity
1	596	80	11	1378	71
2	699	100	12	1453	97
3	736	99	13	1496	76
4	805	40	14	1603	47
5	912	43	15	1702	90
6	1021	99	16	2875	850
7	1077	64	17	2932	78
8	1113	63	18	3031	90
9	1207	86	19	3064	85
10	1275	97	20	3352	100

1,4-Bis(hydroxymethyl)cyclohexane

CH₂OH — structure

CAS No.	– 00105–08–8
PM Ref. No.	– 13390
Restrictions	– none
Formula	– $C_8 H_{16} O_2$
Molecular weight	– 144.22
Alternative names	– 1,4–Cyclohexane dimethanol

Physical Characteristics – Clear viscous liquid, bp 283°C.

Handling – Store at room temperature (25°C).
Safety – Irritant.
Availability – Standard sample supplied.

Current uses – Starting substance – condensation reaction with terephthalic acid dimethyl ester to give 1,4–cyclohexylene dimethylene terephthalate – co-polymer with ethylene. Polyethers.

Applications – Co-extruded film, for packaging bakery products. Adhesives. Mould release agents.

Methods of Characterisation – IR
Mass Spectroscopy

GC Retention Index – 1517
(DB5, 3 min at 50°C, rising 20°C/min⁻¹ to 300°C, hold for 20 min.)

Purity – 99%

78

1,4-Bis(hydroxymethyl)cyclohexane

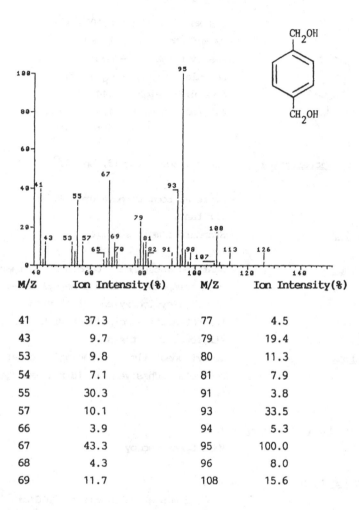

M/Z	Ion Intensity(%)	M/Z	Ion Intensity(%)
41	37.3	77	4.5
43	9.7	79	19.4
53	9.8	80	11.3
54	7.1	81	7.9
55	30.3	91	3.8
57	10.1	93	33.5
66	3.9	94	5.3
67	43.3	95	100.0
68	4.3	96	8.0
69	11.7	108	15.6

Spectrometer   :Finnigan Mat SSQ 70
Inlet System   :Capillary GC/MS
Source Temperature :150°C
Electron Energy   :70 eV
Scan Range   :25–400

1,4-Bis(hydroxymethyl)cyclohexane   Transmission Infra Red

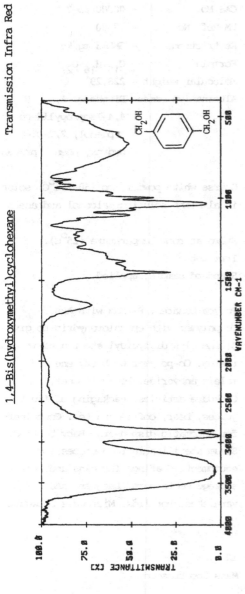

Spectrometer   : Bruker FT-IR IFS 85
Sample   : Thin film between NaCl windows
Solvents   : Not applicable
Reference   : NaCl window
Scan Time   : 1.4 sec/scan
Number of Scans: 60

Peak Table

No.	Wave No.	Rel. Intensity	No.	Wave No.	Rel. Intensity
1	595	42	10	1230	21
2	663	49	11	1282	25
3	884	16	12	1356	37
4	919	19	13	1381	47
5	967	40	14	1448	72
6	996	63	15	1650	28
7	1033	93	16	2854	98
8	1102	49	17	2919	100
9	1195	23	18	3360	98

80

2,2-Bis(4-hydroxyphenyl)propane

CAS No.	– 00080-05-7
PM Ref. No.	– 13480
Restrictions	– SML=3 mg/kg
Formula	– $C_{15} H_{16} O_2$
Molecular weight	– 228.29
Alternative names	– Bisphenol A, 4,4-Isopropylidene-diphenol, 2,2-Di-hydroxy-phenyl propane.

Physical Characteristics – Coarse white powder, mp 150-155°C, soluble in alkaline solution, alcohol and acetone.

Handling – Store at room temperature (25°C).

Safety – Irritant.

Availability – Standard sample supplied.

Current uses – Polycarbonates. Blends with ABS. Co-polymer with epichlorohydrin to give Bisphenol A diglycidyl ether used in epoxy resins. Co-polymer with styrene and maleic anhydride. Polysulphones.

Applications – Cookware articles. Packaging of fruit juices, beer, coffee and tea. Containers for automatic dispensers. Baby bottles. Steam sterilisable food processing equipment. Coatings for cans and bulk storage containers. Lacquers and varnishes. Pot lids. Microwave cookware.

Methods of Characterisation – IR
Mass Spectroscopy

GC Retention Index   – 2237
                          (DB5, 3 min at 50°C, rising 20°C/min^{-1} to
                          300°C, hold for 20 min.)

Purity   – 99%

Analytical methods   – Extraction of aqueous simulants on to
                          solid–phase cartridge, elution with
                          methanol/dichloromethane, concentration
                          then HPLC (UV detection). In wines similar
                          extraction and analysis by HPLC
                          (fluorescence detection).

References   – Method under development (TNO, Zeist, NL).
                          J. Chromatogr., 1986, 360, 266–270
                          J. Chromatogr., 1987, 407, 384–388

2,2-Bis(4-hydroxyphenyl)propane

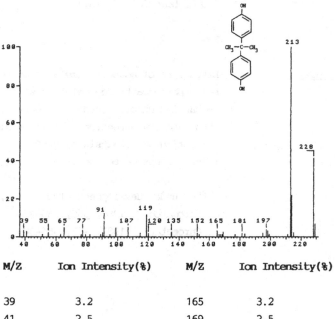

M/Z	Ion Intensity(%)	M/Z	Ion Intensity(%)
39	3.2	165	3.2
41	2.5	169	2.5
55	2.1	195	1.8
65	4.8	197	3.5
77	3.2	198	2.9
91	8.2	213	100.0
99	4.6	214	21.6
107	4.9	215	2.1
119	11.7	228	41.2
135	2.4	229	6.7

Spectrometer   :Finnigan Mat SSQ 70
Inlet System   :Capillary GC/MS
Source Temperature:150°C
Electron Energy   :70 eV
Scan Range   :25–400

2.2-Bis(4-hydroxyphenyl)propane   Transmission Infra Red

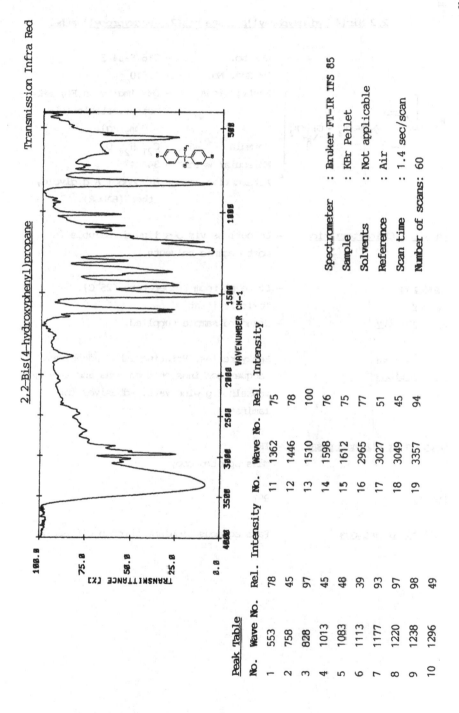

Spectrometer	: Bruker FT-IR IFS 85
Sample	: KBr Pellet
Solvents	: Not applicable
Reference	: Air
Scan time	: 1.4 sec/scan
Number of scans: 60	

Peak Table

No.	Wave No.	Rel. Intensity	No.	Wave No.	Rel. Intensity
1	553	78	11	1362	75
2	758	45	12	1446	78
3	828	97	13	1510	100
4	1013	45	14	1598	76
5	1083	48	15	1612	75
6	1113	39	16	2965	77
7	1177	93	17	3027	51
8	1220	97	18	3049	45
9	1238	98	19	3357	94
10	1296	49			

2,2-Bis(4-hydroxyphenyl)propane bis(2,3-epoxypropyl) ether

CAS No.   — 01675-54-3

PM Ref. No.   — 13510

Restrictions   — QM= 1mg/kg in FP, SML=
not detectable (DL=
0.02mg/kg).

Formula   — $C_{21} H_{24} O_4$

Molecular weight — 340.42

Alternative names— Bisphenol A diglycidyl
ether (BADGE).

| Physical Characteristics | — Colourless viscous liquid. Soluble in most organic solvents. |

Physical Characteristics — Colourless viscous liquid. Soluble in
most organic solvents.

Handling — Store at room temperature (25°C).

Safety — Toxic/Irritant.

Availability — Standard sample supplied.

Current uses — Epoxy resins. Printing inks. Adhesives.

Applications — Lacquer coatings on food cans and storage
vessels e.g wine vats. Adhesives for
laminates.

Methods of Characterisation — IR
Mass Spectroscopy

Purity — 99%

Analytical methods — Pass aqueous simulant through C18 Sep-Pak cartridge, elute with methanol, concentrate and analyse by HPLC (acetonitrile/water gradient) using fluorescence detection. Extract olive oil simulant with water and treat as above.

References

- Method under development (University of Santiago de Compostela, Santiago, E). J. Assoc. Off. Anal. Chem., 1991, 74, 925-928.

J. Chromatogr., 1986, 360, 266-270

2,2-Bis(4-hydroxyphenyl)propane bis(2,3-epoxypropyl)ether

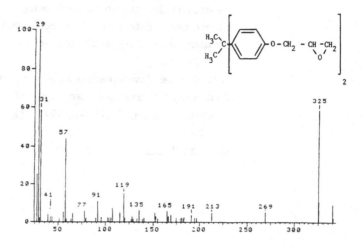

M/Z	Ion Intensity(%)	M/Z	Ion Intensity(%)
27	25.2	115	4.8
29	100.0	119	15.0
31	58.7	135	6.2
41	7.1	152	4.8
55	5.0	165	5.8
57	44.2	213	4.8
65	4.4	269	5.2
77	5.8	325	58.8
91	11.2	326	10.5
107	7.5	340	9.3

Spectrometer   :Finnigan Mat SSQ 70
Inlet System   :Capillary GC/MS
Source Temperature:150°C
Electron Energy   :70 eV
Scan Range   :25-400

2,2-Bis(4-hydroxyphenyl)propane bis(2,3-epoxypropyl) ether   Transmission Infra Red

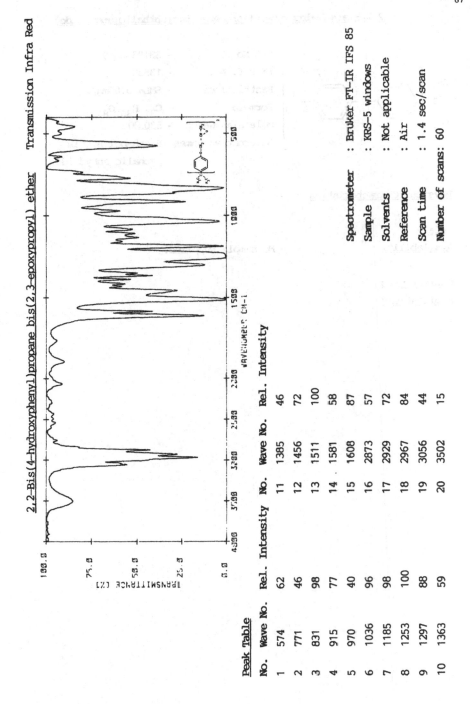

Spectrometer	: Bruker FT-IR IFS 85
Sample	: KRS-5 windows
Solvents	: Not applicable
Reference	: Air
Scan time	: 1.4 sec/scan
Number of scans: 60	

Peak Table

No.	Wave No.	Rel. Intensity	No.	Wave No.	Rel. Intensity
1	574	62	11	1385	46
2	771	46	12	1456	72
3	831	98	13	1511	100
4	915	77	14	1581	58
5	970	40	15	1608	87
6	1036	96	16	2873	57
7	1185	98	17	2929	72
8	1253	100	18	2967	84
9	1297	88	19	3056	44
10	1363	59	20	3502	15

88

2,2-Bis(4-hydroxyphenyl)propane, bis(phthalic anhydride)

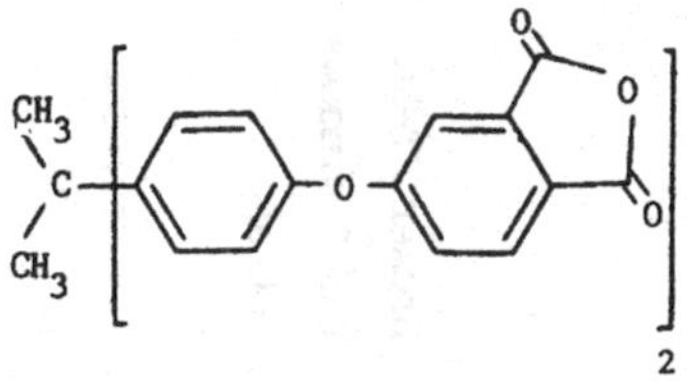

CAS No.	– 38103–06–9
PM Ref. No.	– 13530
Restrictions	– SML= 0.05mg/kg
Formula	– $C_{31} H_{30} O_8$
Molecular weight	– 530.00
Alternative names	– Bisphenol A, bis phthalic anhydride.

Physical Characteristics   –

Safety   –
Availability   – No sample supplied.

Current uses   –
Applications   –

3,3-Bis(3-methyl-4-hydroxyphenyl)-2-indolinone

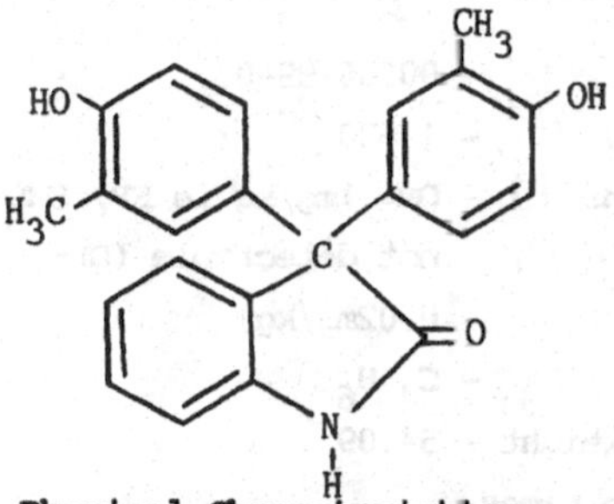

CAS No.   – 47465-97-4

PM Ref. No.   – 13600

Restrictions   – SML= 1.8mg/kg

Formula   – $C_{22}H_{19}NO_3$

Molecular weight – 345.40

Alternative names–

Physical Characteristics   –

Safety   –

Availability   – No sample supplied.

Current uses   –

Applications   –

Analytical methods   –

References   – Method under development (PIRA International, Leatherhead, UK).

1,3-Butadiene

$$CH_2=CH-CH=CH_2$$

CAS No.	– 00106-99-0
PM Ref. No.	– 13630
Restrictions	– QM= 1mg/kg in FP, SML= not detectable (DL= 0.02mg/kg)
Formula	– C_4H_6
Molecular weight	– 54.09
Alternative names–	

Physical Characteristics – Colourless gas, mp –109°C, bp –4.5°C. Soluble in alcohol, dimethyl acetamide and most organic solvents. Inhibited with p-tert-butylcatechol.

Handling – Store at room temperature (25°C).

Safety – Suspect carcinogen/Flammable.

Availability – Standard sample supplied, as a solution in dimethyl acetamide at a concentration of 10mg/ml.

Current uses – Polybutadiene, co-polymer with methacrylonitrile. Used to make co-polymer with acrylonitrile and styrene (ABS, BS). As co-polymers with vinyl chloride.

Applications – Coatings. Margarine tubs, trays, boxes. Refrigerator linings, kitchen appliances. Carbonated beverage bottles. Adhesives.

Methods of Characterisation – Mass Spectroscopy

Purity – 99%

Analytical methods — Dissolution of polymer in dimethylacetamide, followed by headspace analysis (equilibration at 70-80°C) using FID or GC/MS (m/z = 53 and 54). Direct headspace analysis of food simulants or foods such as soft margarine.

References — Draft CEN method (Fraunhofer Inst. Munich, D).
J. Chromatogr., (1984) 294, 427-430.
De Ware(n)-Chemicus (1985) 15, 136-145.
J. Assoc. Off. Anal. Chem., (1987) 70, 18-21.

1,3-Butadiene

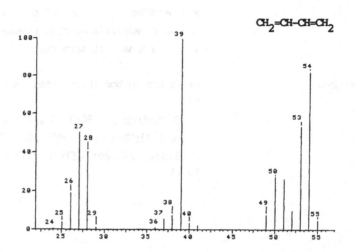

$CH_2=CH-CH=CH_2$

M/Z	Ion Intensity(%)	M/Z	Ion Intensity(%)
24	0.3	40	3.2
25	3.4	41	2.3
26	19.0	49	7.7
27	50.1	50	27.2
28	40.2	51	26.5
29	1.3	52	9.8
36	1.0	53	53.3
37	5.4	54	82.1
38	7.4	55	4.5
39	100.0		

Spectrometer   :Finnigan Mat SSQ 70
Inlet System   :Capillary GC/MS
Source Temperature:150°C
Electron Energy   :70 eV
Scan Range   :25-400

1,3-Butanediol

OH
|
$CH_3-CH-CH_2-CH_2OH$

CAS No.	– 00107-88-0
PM Ref. No.	– 13690
Restrictions	– none
Formula	– $C_4 H_{10} O_2$
Molecular weight	– 90.12
Alternative names	– 1,3-Butylene glycol.

Physical Characteristics – Colourless liquid, mp $-77^{o}C$, bp $203-204^{o}C$.

Handling – Store at room temperature ($25^{o}C$).
Hygroscopic.

Safety – Irritant.

Availability – Standard sample supplied.

Current uses – Used in unsaturated polyesters to impart rigidity and slow cure properties. Used in polyvinyl chloride resins.

Applications –

Methods of Characterisation – IR
Mass Spectroscopy

Purity – 99%

1,3-Butanediol

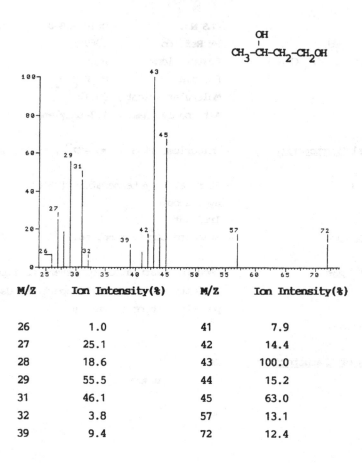

M/Z	Ion Intensity(%)	M/Z	Ion Intensity(%)
26	1.0	41	7.9
27	25.1	42	14.4
28	18.6	43	100.0
29	55.5	44	15.2
31	46.1	45	63.0
32	3.8	57	13.1
39	9.4	72	12.4

Spectrometer   :Finnigan Mat SSQ 70
Inlet System   :Capillary GC/MS
Source Temperature:150°C
Electron Energy   :70 eV
Scan Range   :25-400

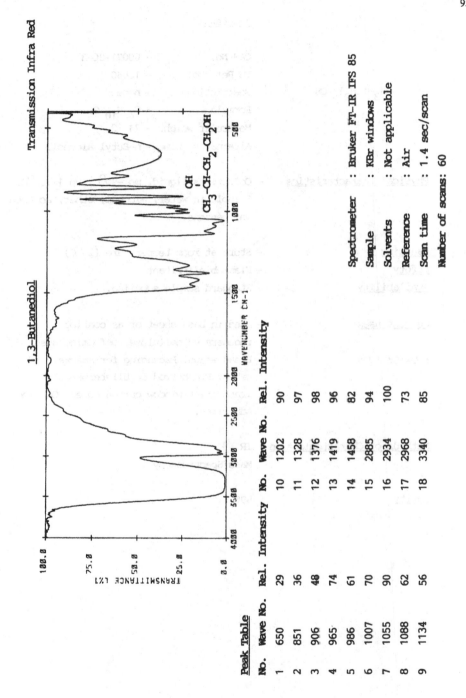

1,3-Butanediol

Transmission Infra Red

$CH_3-CH-CH_2-CH_2OH$
|
OH

TRANSMITTANCE [%]

WAVENUMBER CM-1

Spectrometer	: Bruker FT-IR IFS 85
Sample	: KBr windows
Solvents	: Not applicable
Reference	: Air
Scan time	: 1.4 sec/scan
Number of scans: 60

Peak Table

No.	Wave No.	Rel. Intensity	No.	Wave No.	Rel. Intensity
1	650	29	10	1202	90
2	851	36	11	1328	97
3	906	48	12	1376	98
4	965	74	13	1419	96
5	986	61	14	1458	82
6	1007	70	15	2885	94
7	1055	90	16	2934	100
8	1088	62	17	2968	73
9	1134	56	18	3340	85

1-Butanol

$CH_3-(CH_2)_3-OH$

CAS No.	– 00071–36–3
PM Ref. No.	– 13840
Restrictions	– none
Formula	– $C_4 H_{10} O$
Molecular weight	– 74.12
Alternative names	– n–Butyl alcohol.

Physical Characteristics – Colourless liquid, mp –90°C, bp 117.7^{O}C. Soluble in water, alcohol, ether, acetone and benzene.

Handling – Store at room temperature (25°C).
Safety – Flammable/Irritant.
Availability – Standard sample supplied.

Current uses – Used in base sheet or as coating in regenerated cellulose. Defoaming agent.

Applications – Bread wraps. Packaging for snacks, and cakes. Fresh Produce blisterpacks. Transparent window carton boxes. Coatings. Adhesives.

Methods of Characterisation – IR
Mass Spectroscopy

Purity – 99%

1-Butanol

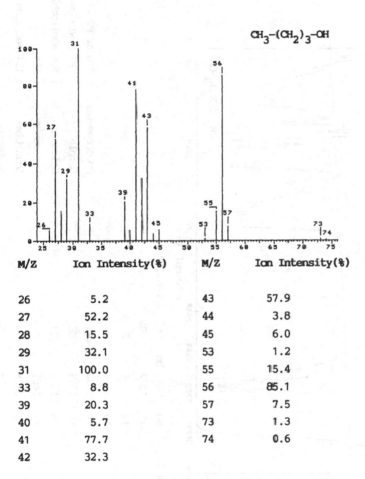

$$CH_3-(CH_2)_3-OH$$

M/Z	Ion Intensity(%)	M/Z	Ion Intensity(%)
26	5.2	43	57.9
27	52.2	44	3.8
28	15.5	45	6.0
29	32.1	53	1.2
31	100.0	55	15.4
33	8.8	56	85.1
39	20.3	57	7.5
40	5.7	73	1.3
41	77.7	74	0.6
42	32.3		

Spectrometer   :Finnigan Mat SSQ 70
Inlet System   :Capillary GC/MS
Source Temperature:150°C
Electron Energy   :70 eV
Scan Range   :25–400

Transmission Infra Red

1-Butanol

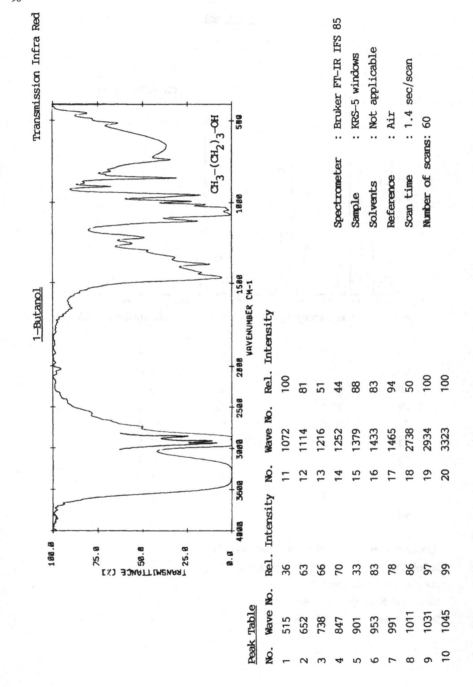

$CH_3-(CH_2)_3-OH$

Spectrometer	: Bruker FT-IR IFS 85
Sample	: KRS-5 windows
Solvents	: Not applicable
Reference	: Air
Scan time	: 1.4 sec/scan
Number of scans: 60	

Peak Table

No.	Wave No.	Rel. Intensity	No.	Wave No.	Rel. Intensity
1	515	36	11	1072	100
2	652	63	12	1114	81
3	738	66	13	1216	51
4	847	70	14	1252	44
5	901	33	15	1379	88
6	953	83	16	1433	83
7	991	78	17	1465	94
8	1011	86	18	2738	50
9	1031	97	19	2934	100
10	1045	99	20	3323	100

1-Butene

$CH_3-CH_2-CH=CH_2$

CAS No.   – 00106–98–9
PM Ref. No.   – 13870
Restrictions   – none
Formula   – $C_4 H_8$
Molecular weight – 56.11
Alternative names– Butylene;
                   ethylethylene.

Physical Characteristics   – Colourless gas, mp $-185^{\circ}C$, bp $-6.5^{\circ}C$. Soluble in most organic solvents.

Safety   – Flammable.

Availability   – No sample supplied.

Current uses   – To make polybutene (polybutylene). A co-monomer for linear low density polyethylene. A co-monomer for high density polyethylene. Co-polymer with terephthalic acid (polybutyl terephthalate), laminated with polystyrene. Used to make butyl rubber and butadiene.

Applications   – Film for pre-packed and frozen foods. Heavy duty films for shipping bags. Lubricant. Part of laminate with paper and aluminium. Blow moulded to form storage containers. For bottles, bags, tubs, trays, crates and utensils. Tubes used in potable water services. Used to seal yoghurt pots. Chain inhibitors.

2-Butene

$CH_3-CH=CH-CH_3$

CAS No.	– 00107–01–7
PM Ref. No.	– 13900
Restrictions	– none
Formula	– $C_4 H_8$
Molecular weight	– 56.11
Alternative names	– 2–Butylene;
	1,2–dimethyl ethylene.

Physical Characteristics — Colourless gas, mp –139.3^{o}C (cis form), & mp –105.8^{o}C (trans form), bp 3.7^{o}C (cis form), bp 1°C (trans form).

Safety — Flammable.

Availability — No sample supplied.

Current uses — Polybutene (polybutylenes). Used to synthesise butyl rubber, butadiene and cresols.

Applications — Lubricants, paper applications. Chain inhibitors.

Butyraldehyde

CH₃CH₂CH₂CHO

CAS No.	– 00123-72-8
PM Ref. No.	– 14110
Restrictions	– none
Formula	– $C_4 H_8 O$
Molecular weight	– 72.11
Alternative names	– Butanal, n-butyric aldehyde.

$CH_3CH_2CH_2CHO$

Physical Characteristics – Colourless liquid, mp $-99^{\circ}C$, bp $74.8^{\circ}C$. Miscible with ethanol, ether, acetone, and toluene. Soluble in water.

Handling – Store at room temperature ($25^{\circ}C$).

Safety – Flammable/Irritant.

Availability – Standard sample supplied.

Current uses – Inhibitor for urea-formaldehyde resins. Chain regulator for vinyl chloride polymers, and grafting of vinyl chloride on to ethylene polymers. Cross-linking agent for resorcinol (1,3-Dihyroxybenzene) co-polymers.

Applications – Repeat-use articles and picnic wear. Polyvinyl chloride films and tubs.

Methods of Characterisation – IR
Mass Spectroscopy

Purity – 99%

Butyraldehyde

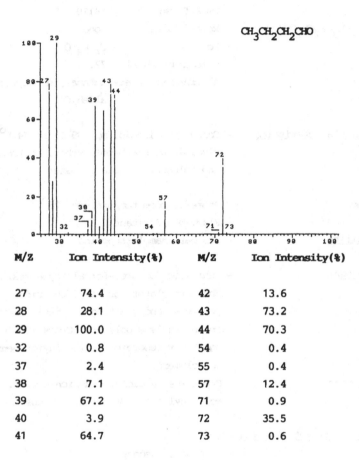

$CH_3CH_2CH_2CHO$

M/Z	Ion Intensity(%)	M/Z	Ion Intensity(%)
27	74.4	42	13.6
28	28.1	43	73.2
29	100.0	44	70.3
32	0.8	54	0.4
37	2.4	55	0.4
38	7.1	57	12.4
39	67.2	71	0.9
40	3.9	72	35.5
41	64.7	73	0.6

Spectrometer   :Finnigan Mat SSQ 70
Inlet System   :Capillary GC/MS
Source Temperature:150°C
Electron Energy   :70 eV
Scan Range   :25-400

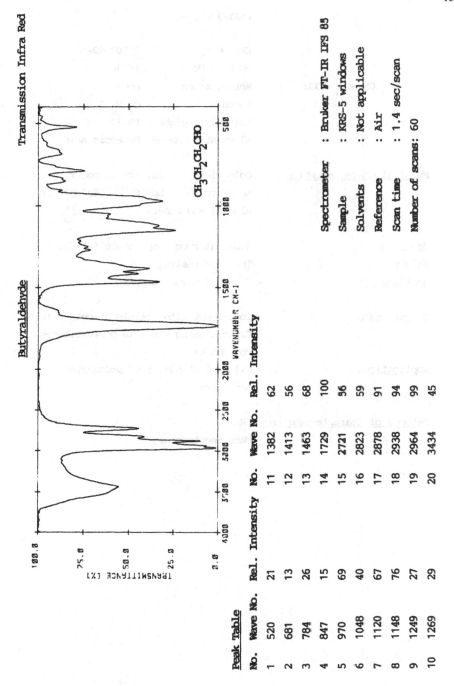

Butyraldehyde   Transmission Infra Red

$CH_3CH_2CH_2CHO$

Spectrometer	:	Bruker FT-IR IFS 85
Sample	:	KRS-5 windows
Solvents	:	Not applicable
Reference	:	Air
Scan time	:	1.4 sec/scan
Number of scans: 60		

Peak Table

No.	Wave No.	Rel. Intensity	No.	Wave No.	Rel. Intensity
1	520	21	11	1382	62
2	681	13	12	1413	56
3	784	26	13	1463	68
4	847	15	14	1729	100
5	970	69	15	2721	56
6	1048	40	16	2823	59
7	1120	67	17	2878	91
8	1148	76	18	2938	94
9	1249	27	19	2964	99
10	1269	29	20	3434	45

Butyric acid

CAS No.	- 00107-92-6
PM Ref. No.	- 14140
Restrictions	- none
Formula	- $C_4 H_8 O_2$
Molecular weight	- 88.11
Alternative names	- Butanoic acid.

$CH_3-CH_2-CH_2-COOH$

Physical Characteristics - Colourless liquid, strong odour,
mp -7 to -5°C, bp 162°C, Soluble in
alcohol and ether.

Handling   - Store at room temperature (25°C).
Safety   - Toxic/Corrosive.
Availability   - Standard sample supplied.

Current uses   - Co-polymer with cellulose acetate (CAB).
Cross-linking agent in polyurethanes.
Polyamides.
Applications   - Films. Coatings. Food processing
machinery.

Methods of Characterisation - IR
Mass Spectroscopy

Purity   - 99%

Butyric acid

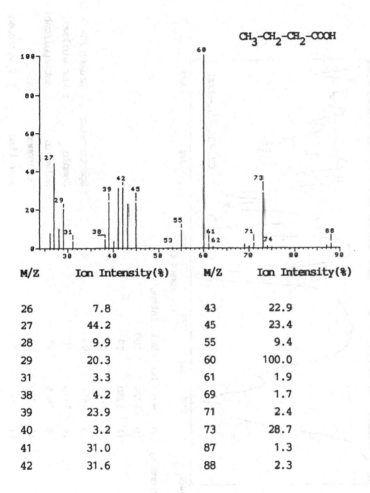

$CH_3-CH_2-CH_2-COOH$

M/Z	Ion Intensity(%)	M/Z	Ion Intensity(%)
26	7.8	43	22.9
27	44.2	45	23.4
28	9.9	55	9.4
29	20.3	60	100.0
31	3.3	61	1.9
38	4.2	69	1.7
39	23.9	71	2.4
40	3.2	73	28.7
41	31.0	87	1.3
42	31.6	88	2.3

Spectrometer   :Finnigan Mat SSQ 70
Inlet System   :Capillary GC/MS
Source Temperature:150°C
Electron Energy   :70 eV
Scan Range   :25–400

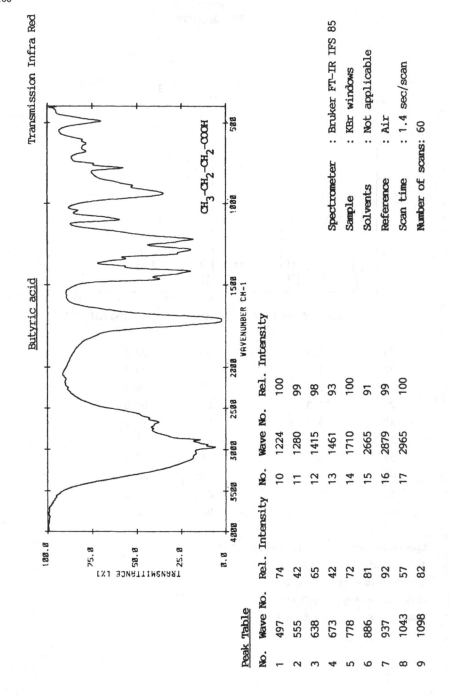

Butyric acid

Transmission Infra Red

$CH_3-CH_2-CH_2-COOH$

WAVENUMBER CM-1

TRANSMITTANCE [%]

Peak Table

No.	Wave No.	Rel. Intensity	No.	Wave No.	Rel. Intensity
1	497	74	10	1224	100
2	555	42	11	1280	99
3	638	65	12	1415	98
4	673	42	13	1461	93
5	778	72	14	1710	100
6	886	81	15	2665	91
7	937	92	16	2879	99
8	1043	57	17	2965	100
9	1098	82			

Spectrometer : Bruker FT-IR IFS 85
Sample   : KBr windows
Solvents   : Not applicable
Reference   : Air
Scan time   : 1.4 sec/scan
Number of scans: 60

Butyric anhydride

$(CH_3-CH_2-CH_2-CO)_2-O$

CAS No.	— 00106–31–0
PM Ref. No.	— 14170
Restrictions	— none
Formula	— $C_8 H_{14} O_3$
Molecular weight	— 158.20
Alternative names—	

Physical Characteristics — Pale amber, pungent liquid, mp -75^OC, bp 199–201°C. Soluble in ether.

Handling — Store at room temperature (25°C). Moisture sensitive.

Safety — Corrosive.

Availability — Standard sample supplied.

Current uses — Used as a starting substance in the synthesis of cellulose acetate butyrate.

Applications — Shock resistant plastics. Films.

Methods of Characterisation — IR
Mass Spectroscopy

Purity — 99%

Butyric anhydride

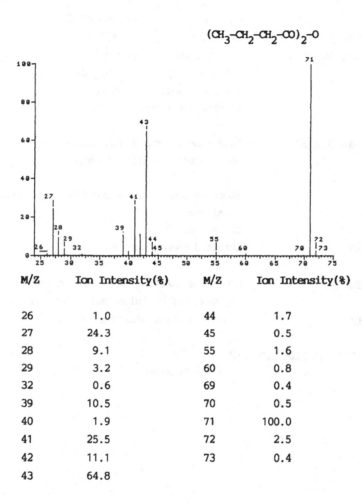

$$(CH_3-CH_2-CH_2-CO)_2-O$$

M/Z	Ion Intensity(%)	M/Z	Ion Intensity(%)
26	1.0	44	1.7
27	24.3	45	0.5
28	9.1	55	1.6
29	3.2	60	0.8
32	0.6	69	0.4
39	10.5	70	0.5
40	1.9	71	100.0
41	25.5	72	2.5
42	11.1	73	0.4
43	64.8		

Spectrometer   :Finnigan Mat SSQ 70
Inlet System   :Capillary GC/MS
Source Temperature:150°C
Electron Energy   :70 eV
Scan Range   :25–400

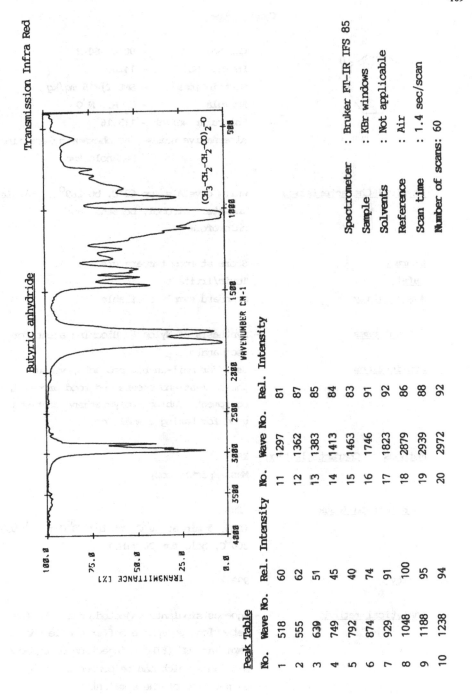

Butyric anhydride

Transmission Infra Red

$(CH_3-CH_2-CH_2-CO)_2-O$

Spectrometer	: Bruker FT-IR IFS 85
Sample	: KBr windows
Solvents	: Not applicable
Reference	: Air
Scan time	: 1.4 sec/scan
Number of scans: 60	

Peak Table

No.	Wave No.	Rel. Intensity	No.	Wave No.	Rel. Intensity
1	518	60	11	1297	81
2	555	62	12	1362	87
3	639	51	13	1383	85
4	749	45	14	1413	84
5	792	40	15	1463	83
6	874	74	16	1746	91
7	929	91	17	1823	92
8	1048	100	18	2879	86
9	1188	95	19	2939	88
10	1238	94	20	2972	92

Caprolactam

CAS No.	– 00105-60-2
PM Ref. No.	– 14200
Restrictions	– SML(T)=15 mg/kg
Formula	– $C_6 H_{11} N O$
Molecular weight	– 113.16
Alternative names	– 2-Oxohexamethyleneimine
	Hexanolactam

Physical Characteristics – White crystals, mp 69°C, bp 269°C. Soluble in water, alcohol, benzene, and chloroform.

Storage – Store at room temperature (25°C).

Safety – Toxic/Irritant.

Availability – Standard sample available.

Current uses – Synthesis of Nylon-6. Blocking agent for isocyanates.

Applications – Used for boil-in-bag products, vacuum packed meat and cheese and food processing equipment. Tubing. Polyurethane coatings used for baking enamels etc.

Methods of Characterisation – IR

Mass Spectroscopy

GC Retention Index – 1294
(DB5, 3 min at 50°C, rising 20°C/min⁻¹ to 300°C, hold for 20 min.)

Purity – 99%

Analytical methods – Aqueous simulants injected onto HPLC (UV detection)-phosphate buffer mobile phase. Capillary GC (FID) - injection of aqueous simulants which can be concentrated by evaporation of the simulant.

References   – Method under development (Fraunhofer,
                                Munich, D.)

112

Caprolactam

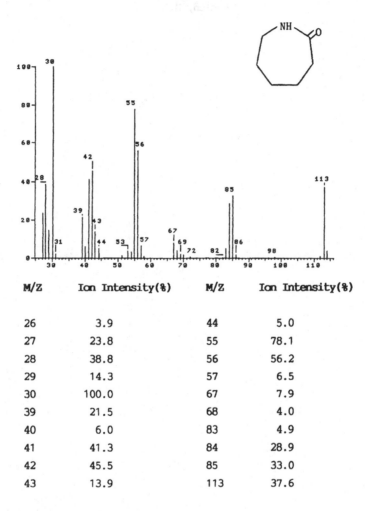

M/Z	Ion Intensity(%)	M/Z	Ion Intensity(%)
26	3.9	44	5.0
27	23.8	55	78.1
28	38.8	56	56.2
29	14.3	57	6.5
30	100.0	67	7.9
39	21.5	68	4.0
40	6.0	83	4.9
41	41.3	84	28.9
42	45.5	85	33.0
43	13.9	113	37.6

Spectrometer   :Finnigan Mat SSQ 70
Inlet System   :Capillary GC/MS
Source Temperature:150°C
Electron Energy   :70 eV
Scan Range   :25-400

113

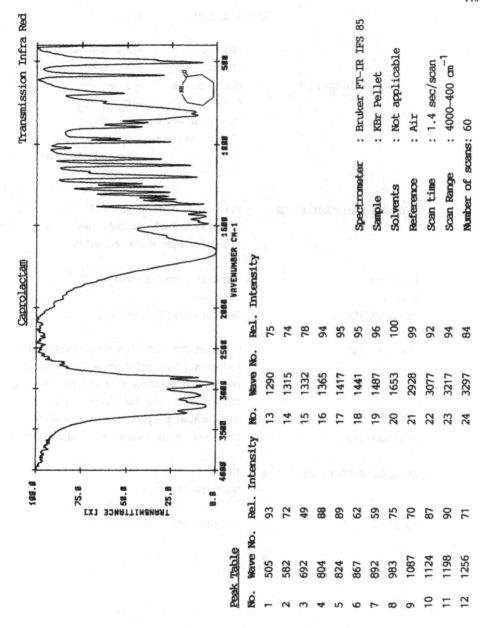

Transmission Infra Red

Caprolactam

Spectrometer	: Bruker FT-IR IFS 85
Sample	: KBr Pellet
Solvents	: Not applicable
Reference	: Air
Scan time	: 1.4 sec/scan
Scan Range	: 4000–400 cm^{-1}
Number of scans: 60	

Peak Table

No.	Wave No.	Rel. Intensity	No.	Wave No.	Rel. Intensity
1	505	93	13	1290	75
2	582	72	14	1315	74
3	692	49	15	1332	78
4	804	88	16	1365	94
5	824	89	17	1417	95
6	867	62	18	1441	95
7	892	59	19	1487	96
8	983	75	20	1653	100
9	1087	70	21	2928	99
10	1124	87	22	3077	92
11	1198	90	23	3217	94
12	1256	71	24	3297	84

114

Caprylic acid

$CH_3-(CH_2)_6-COOH$

CAS No.	– 00124–07–2
PM Ref. No.	– 14320
Restrictions	– none
Formula	– $C_8 H_{16} O_2$
Molecular weight	– 144.22
Alternative names	– Octanoic acid, heptane–1–carboxylic acid.

Physical Characteristics – Pale amber liquid, mp 16.7°C, bp 239°C. Slightly soluble in hot water. Soluble in alcohol, chloroform and ether.

Handling – Store at room temperature (25°C).
Safety – Irritant.
Availability – Standard sample supplied.

Current uses – Catalyst for addition polymerisation of hydroxystyrene polymers with phenolic resins. In formaldehyde resins. Coating filler particles for mixing with unsaturated polyester resins and PVC.
Applications – Coatings. Picnic-ware. Film and sheeting.

Methods of Characterisation – IR
Mass Spectroscopy

Purity – 79% with minor impurities.

Caprylic acid

$CH_3-(CH_2)_6-COOH$

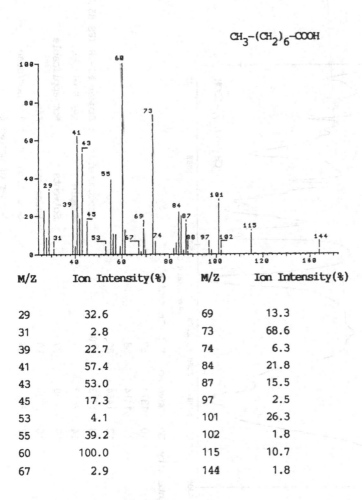

M/Z	Ion Intensity(%)	M/Z	Ion Intensity(%)
29	32.6	69	13.3
31	2.8	73	68.6
39	22.7	74	6.3
41	57.4	84	21.8
43	53.0	87	15.5
45	17.3	97	2.5
53	4.1	101	26.3
55	39.2	102	1.8
60	100.0	115	10.7
67	2.9	144	1.8

Spectrometer   :Finnigan Mat SSQ 70
Inlet System   :Capillary GC/MS
Source Temperature:150°C
Electron Energy   :70 eV
Scan Range   :25-400

116

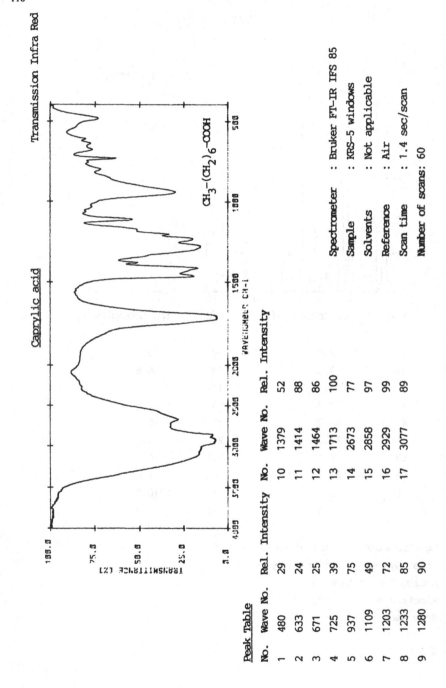

Caprylic acid

Transmission Infra Red

$CH_3-(CH_2)_6-COOH$

Spectrometer : Bruker FT-IR IFS 85

Sample   : KRS-5 windows

Solvents   : Not applicable

Reference   : Air

Scan time   : 1.4 sec/scan

Number of scans: 60

Peak Table

No.	Wave No.	Rel. Intensity	No.	Wave No.	Rel. Intensity
1	480	29	10	1379	52
2	633	24	11	1414	88
3	671	25	12	1464	86
4	725	39	13	1713	100
5	937	75	14	2673	77
6	1109	49	15	2858	97
7	1203	72	16	2929	99
8	1233	85	17	3077	89
9	1280	90			

Carbon monoxide

CAS No.   — 00630—08—0
PM Ref. No.   — 14350
Restrictions   — none
Formula   — C O
Molecular weight   — 28.01
Alternative names—

Physical Characteristics — Colourless odourless gas, mp -205°C,
bp -191.5°C. Soluble in ethyl acetate, and
acetic acid. Sparingly soluble in water.

Safety   — Toxic/Flammable.
Availability   — No sample supplied.

Current uses   — Starting material for the synthesis of
carbonyl chloride (phosgene). Co—polymers
with ethylene. A reducing agent in the
synthesis of isocyanates. Reacts with
acetylene to yield acrylates and
acrylonitrile. Used in the formation of
synthetic paraffin.

Applications   — Polycarbonates used for tableware, packing
fruit juices, and beer. Containers for
automatic dispensers. Photodegradable
polymers. Coatings.

Carbonyl chloride

CAS No.	– 00075–44–5
PM Ref. No.	– 14380
Restrictions	– QM= 1mg/kg in FP
Formula	– C Cl$_2$ O
Molecular weight	– 98.92
Alternative names	– Phosgene.

Cl
    \C=O
Cl

Physical Characteristics – Colourless gas, mp -127°C, bp 8.2°C.
Soluble in benzene, toluene, and glacial
acetic acid.

Safety – Poison.

Availability – No sample supplied.

Current uses – Used to make isocyanates, and
polyurethanes. As a co-polymer with
Bisphenol A to form polycarbonates.

Applications – Adhesive for laminates and confectionary.
Coatings for cookware. For tableware,
packaging for fruit juices and beer.
Containers for automatic dispensers and
for baby feeding bottles.

Analytical methods – Headspace analysis and capillary GC with
electron capture detection. Alternatively
in solution in methylene chloride reaction
with excess 2-aminophenol (10 min) to form
cyclic derivative which is determined by
GC with nitrogen-selective detection.

References – Method under development (Fraunhofer Inst.
Munich, D).
J. Chromatogr., (1989) 481, 373–379.
Analyst, (1983) 108, 974–983.

Castor oil

CAS No.	– 08001-79-4
PM Ref. No.	– 14410
Restrictions	– none
Formula	– undefined
Molecular weight	– undefined
Alternative names–	

Physical Characteristics – Pale yellow liquid, bp 313°C.

Safety – Irritant.

Availability – No sample supplied.

Current uses – Starting substance for nylon 11 (synthesis of 11-aminoundecanoic acid). Drying oil. Polyurethane synthesis. Polyoxyethyl castor oil used as an emulsifier in nitrocellulose.

Applications – Coatings, filter cloths, brushes, bearings and rollers in conveyer systems. Film or enamel coatings for paper and paperboard. In resins to improve flexibility and colour.

Cellulose

CAS No.	– 09004-34-6
PM Ref. No.	– 14500
Restrictions	– none
Formula	– $(C_6 H_{10} O_5)_n$
Molecular weight	– $(162.14)_n$
Alternative names–	

Physical Characteristics — White powder, mp 260-270°C. Soluble in concentrated zinc chloride and caustic alkali.

Handling — Store at room temperature (25°C).

Safety — Irritant.

Availability — Standard sample supplied.

Current uses — Regenerated cellulose film (cellophane), cellulose acetate, cellulose acetate butyrate, cellulose butyrate, ethyl cellulose.

Applications — Bread wraps. Packaging for snacks, cakes meat and milk. Fresh produce blisterpacks. Transparent windows in carton boxes. Sausage casing.

Methods of Characterisation – IR

Purity — Natural product composition variable.

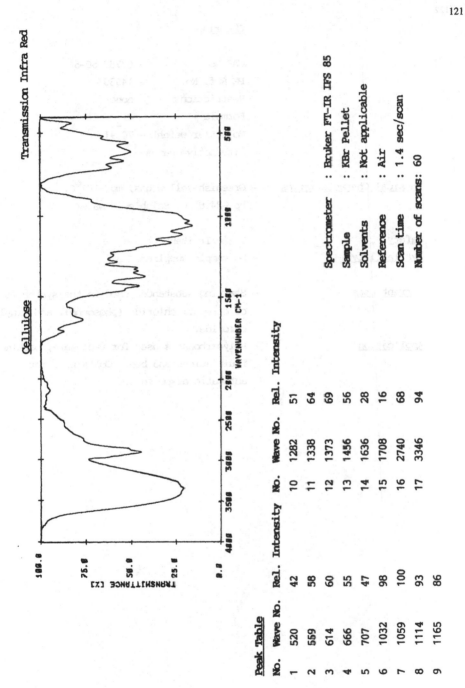

Cellulose   Transmission Infra Red

Peak Table

No.	Wave No.	Rel. Intensity	No.	Wave No.	Rel. Intensity
1	520	42	10	1282	51
2	559	58	11	1338	64
3	614	60	12	1373	69
4	666	55	13	1456	56
5	707	47	14	1636	28
6	1032	98	15	1708	16
7	1059	100	16	2740	68
8	1114	93	17	3346	94
9	1165	86			

Spectrometer : Bruker FT-IR IFS 85
Sample   : KBr Pellet
Solvents   : Not applicable
Reference   : Air
Scan time   : 1.4 sec/scan
Number of scans: 60

Chlorine

CAS No.	– 07782–50–5
PM Ref. No.	– 14530
Restrictions	– none
Formula	– Cl_2
Molecular weight	– 70.91
Alternative names–	

Physical Characteristics — Greenish–yellow gas, mp –101°C, bp –34.05°C. Soluble in water.

Safety — Toxic/Irritant.

Availability — No sample supplied.

Current uses — Starting substance, used in the synthesis of carbonyl chloride (phosgene), and vinyl chloride.

Applications — Polycarbonates used for tableware, packing fruit juices and beer. Containers for automatic dispensers.

Citric acid

HOOC–CH$_2$–C(OH)(COOH)–CH$_2$COOH

CAS No.	– 00077–92–9
PM Ref. No.	– 14680
Restrictions	– none
Formula	– C$_6$ H$_8$ O$_7$
Molecular weight	– 192.13
Alternative names	– 2,–hydroxy–1,2,3–propane tricarboxylic acid.

Physical Characteristics — White crystals, mp 153°C. Soluble in water, ethanol and ether.

Handling — Store at room temperature (25°C).

Safety — Irritant.

Availability — Standard sample supplied.

Current uses — Cross links to cellulose. A nucleating or blowing agent for polymeric foams and for polyolefins. Starting material for esters. Initiator for methyl methacrylate. Polymer with ethylene glycol. Stabiliser for rubber.

Applications — Film and sheeting for bread, cakes, snack foods, meat, cheese and poultry. Adhesive. Food processing machinery.

Methods of Characterisation — IR

Purity — 99%

124

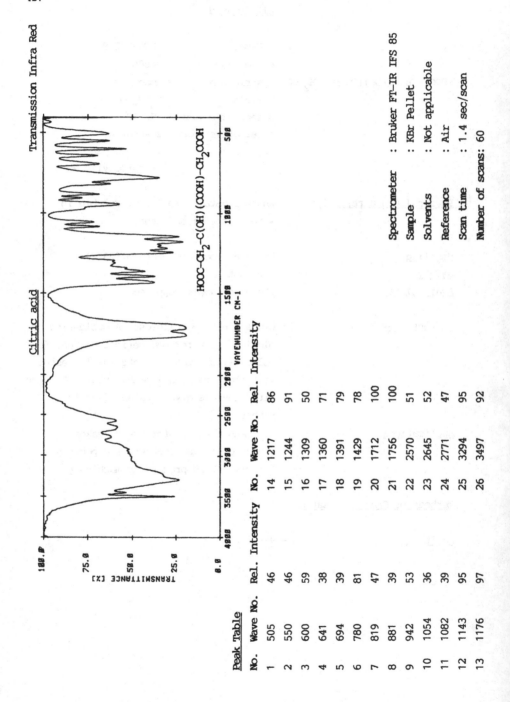

Transmission Infra Red

Citric acid

HOOC-CH$_2$-C(OH)(COOH)-CH$_2$COOH

Spectrometer	:	Bruker FT-IR IFS 85
Sample	:	KBr Pellet
Solvents	:	Not applicable
Reference	:	Air
Scan time	:	1.4 sec/scan
Number of scans:		60

Peak Table

No.	Wave No.	Rel. Intensity	No.	Wave No.	Rel. Intensity
1	505	46	14	1217	86
2	550	46	15	1244	91
3	600	59	16	1309	50
4	641	38	17	1360	71
5	694	39	18	1391	79
6	780	81	19	1429	78
7	819	47	20	1712	100
8	881	39	21	1756	100
9	942	53	22	2570	51
10	1054	36	23	2645	52
11	1082	39	24	2771	47
12	1143	95	25	3294	95
13	1176	97	26	3497	92

m-Cresol

CAS No.	– 00108–39–4
PM Ref. No.	– 14710
Restrictions	– none
Formula	– $C_7 H_8 O$

Molecular weight – 108.14

Alternative names– Meta–Cresol, m–methyl phenol, 3–hydroxytoluene.

Physical Characteristics – Pale amber liquid, mp 11–12°C, bp 197–205°C. Soluble in caustic alkali, ethanol, ether, acetone and benzene. Sparingly soluble in water.

Handling – Store at room temperature (25°C).

Safety – Toxic/Corrosive.

Availability – Standard sample supplied.

Current uses – Co–polymers with formaldehyde to give resins. Polyhydantoins.

Applications – Closures. Can coatings.

Methods of Characterisation – IR

Mass Spectroscopy

Purity – Commercial sample containing substantial amounts of p–cresol, phenol and xylenols.

126

m-Cresol

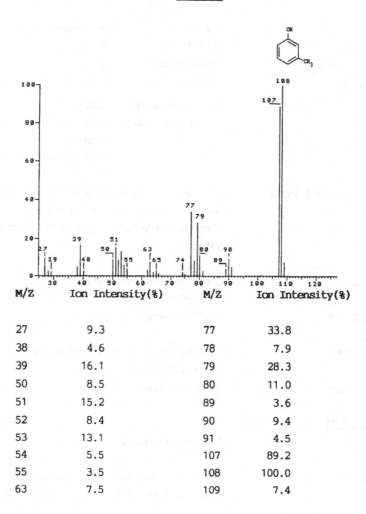

M/Z	Ion Intensity(%)	M/Z	Ion Intensity(%)
27	9.3	77	33.8
38	4.6	78	7.9
39	16.1	79	28.3
50	8.5	80	11.0
51	15.2	89	3.6
52	8.4	90	9.4
53	13.1	91	4.5
54	5.5	107	89.2
55	3.5	108	100.0
63	7.5	109	7.4

Spectrometer   :Finnigan Mat SSQ 70
Inlet System   :Capillary GC/MS
Source Temperature:150°C
Electron Energy   :70 eV
Scan Range   :25–400

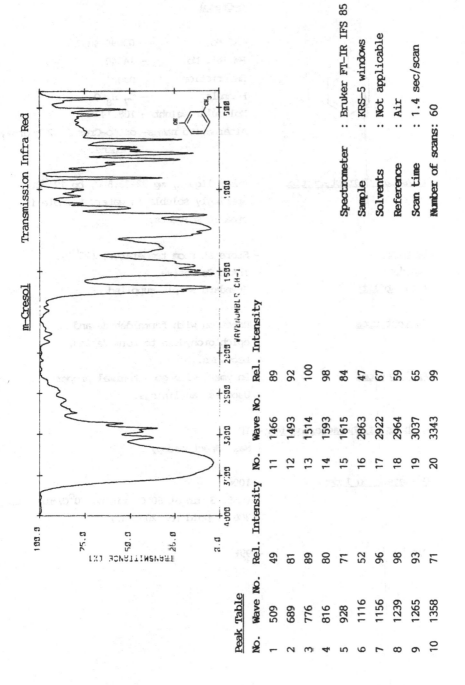

m-Cresol   Transmission Infra Red

Spectrometer : Bruker FT-IR IFS 85
Sample : KRS-5 windows
Solvents : Not applicable
Reference : Air
Scan time : 1.4 sec/scan
Number of scans: 60

Peak Table

No.	Wave No.	Rel. Intensity	No.	Wave No.	Rel. Intensity
1	509	49	11	1466	89
2	689	81	12	1493	92
3	776	89	13	1514	100
4	816	80	14	1593	98
5	928	71	15	1615	84
6	1116	52	16	2863	47
7	1156	96	17	2922	67
8	1239	98	18	2964	59
9	1265	93	19	3037	65
10	1358	71	20	3343	99

o-Cresol

OH
CH₃

CAS No.	– 00095–48–7
PM Ref. No.	– 14740
Restrictions	– none
Formula	– $C_7 H_8 O$
Molecular weight	– 108.14
Alternative names	– ortho–Cresol, 2–hydroxy toluene.

Physical Characteristics – Amber liquid, mp 32–33.5°C, bp 191°C. Sparingly soluble in water. Soluble in organic solvents.

Handling – Store at room temperature (25°C).
Safety – Toxic/Corrosive.
Availability – Standard sample supplied.

Current uses – Reaction with formaldehyde and epichlorohydrin to form resins. Lacquers.
Applications – Polymers with good thermal properties. Used as can linings.

Methods of Characterisation – IR
Mass Spectroscopy

GC Retention Index – 1006
(DB5, 3 min at 50°C, rising 20°C/min⁻¹ to 300°C, hold for 20 min.)

Purity – 99%

o-Cresol

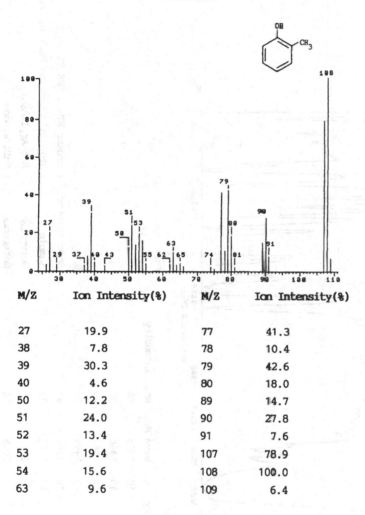

M/Z	Ion Intensity(%)	M/Z	Ion Intensity(%)
27	19.9	77	41.3
38	7.8	78	10.4
39	30.3	79	42.6
40	4.6	80	18.0
50	12.2	89	14.7
51	24.0	90	27.8
52	13.4	91	7.6
53	19.4	107	78.9
54	15.6	108	100.0
63	9.6	109	6.4

Spectrometer   :Finnigan Mat SSQ 70
Inlet System   :Capillary GC/MS
Source Temperature:150°C
Electron Energy   :70 eV
Scan Range   :25-400

130

Transmission Infra Red

o-Cresol

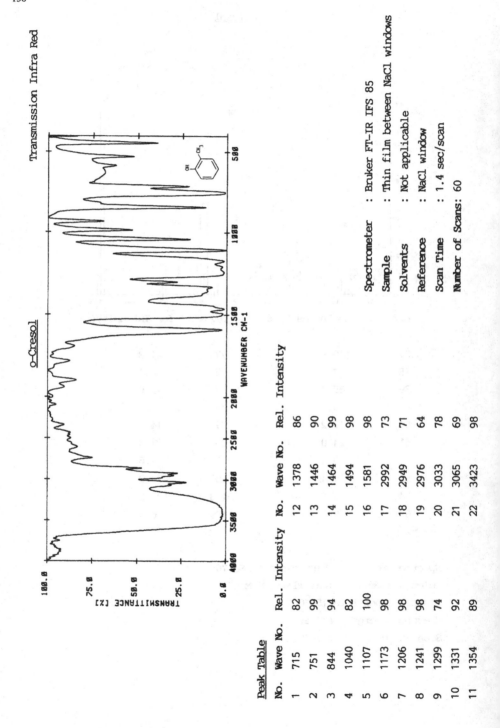

WAVENUMBER CM-1

TRANSMITTANCE [%]

Spectrometer : Bruker FT-IR IFS 85

Sample   : Thin film between NaCl windows

Solvents   : Not applicable

Reference   : NaCl window

Scan Time   : 1.4 sec/scan

Number of Scans: 60

Peak Table

No.	Wave No.	Rel. Intensity	No.	Wave No.	Rel. Intensity
1	715	82	12	1378	86
2	751	99	13	1446	90
3	844	94	14	1464	99
4	1040	82	15	1494	98
5	1107	100	16	1581	98
6	1173	98	17	2992	73
7	1206	98	18	2949	71
8	1241	98	19	2976	64
9	1299	74	20	3033	78
10	1331	92	21	3065	69
11	1354	89	22	3423	98

p-Cresol

OH

CH$_3$

CAS No.	— 00106—44—5
PM Ref. No.	— 14770
Restrictions	— none
Formula	— C$_7$ H$_8$ O
Molecular weight	— 108.14
Alternative names	— 4—Methylphenol, 4—hydroxytoluene.

Physical Characteristics — Colourless liquid, mp 32–34°C, bp 202°C. Sparingly soluble in water. Soluble in aqueous solutions of alkali hydroxides, alcohol, ether, acetone and benzene.

Handling — Store at room temperature (25°C).

Safety — Toxic/Corrosive.

Availability — Standard sample supplied.

Current uses — Co-polymers with formaldehyde to give resins.

Applications — Used as can coatings.

Methods of Characterisation — IR
Mass Spectroscopy

Purity — 94% (2—methylphenyl acetate, impurity).

132

p-Cresol

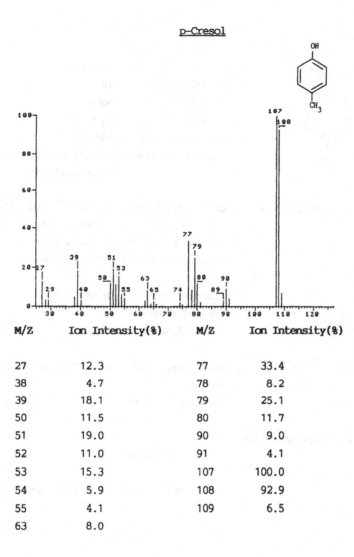

M/Z	Ion Intensity(%)	M/Z	Ion Intensity(%)
27	12.3	77	33.4
38	4.7	78	8.2
39	18.1	79	25.1
50	11.5	80	11.7
51	19.0	90	9.0
52	11.0	91	4.1
53	15.3	107	100.0
54	5.9	108	92.9
55	4.1	109	6.5
63	8.0		

Spectrometer   :Finnigan Mat SSQ 70
Inlet System   :Capillary GC/MS
Source Temperature:150°C
Electron Energy   :70 eV
Scan Range   :25–400

p-Cresol

Transmission Infra Red

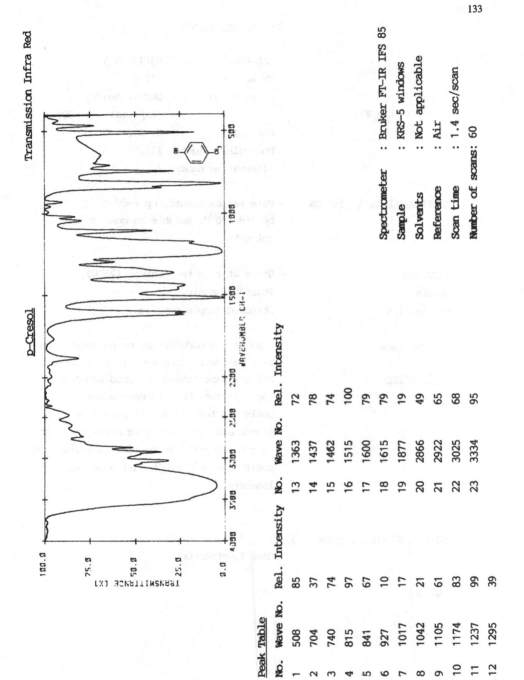

Spectrometer	: Bruker FT-IR IFS 85
Sample	: KRS-5 windows
Solvents	: Not applicable
Reference	: Air
Scan time	: 1.4 sec/scan
Number of scans: 60	

Peak Table

No.	Wave No.	Rel. Intensity	No.	Wave No.	Rel. Intensity
1	508	85	13	1363	72
2	704	37	14	1437	78
3	740	74	15	1462	74
4	815	97	16	1515	100
5	841	67	17	1600	79
6	927	10	18	1615	79
7	1017	17	19	1877	19
8	1042	21	20	2866	49
9	1105	61	21	2922	65
10	1174	83	22	3025	68
11	1237	99	23	3334	95
12	1295	39			

Cyclohexyl Isocyanate

C_6H_{11}-NCO

CAS No.	– 03173–53–3
PM Ref. No.	– 14950
Restrictions	– QM(T)= 1mg/kg
	(expressed as NCO).
Formula	– $C_7 H_{11} N O$
Molecular weight	– 125.17
Alternative names–	

Physical Characteristics – Pale yellow liquid, mp <–80°C,
bp 168–170°C. Soluble in most organic
solvents.

Handling – Store at room temperature (25°C).

Safety – Toxic/Flammable.

Availability – Standard sample supplied.

Current uses – Used in the manufacture of polyurethanes.
As a polymerisation aid for polyamides.

Applications – Polyurethane tubing for food manufacturing
applications. Used to make adhesives in
seals for thin films, in polyester
paperboard laminates (e.g susceptors) and
in multi-layer high barrier plastics (e.g
shelf stables) and 'boil-in-the-bag'
laminates.

Methods of Characterisation – IR
Mass Spectroscopy

Purity – 98%

Analytical methods — Isocyanates in materials and articles are
analysed by solvent extraction with
ethanol in toluene with concurrent
urethane derivative formation, clean—up by
liquid/liquid partition and solid phase
cartridge chromatography and determination
by capillary GC with nitrogen specific
detection. Phenyl isocyanate and
1,4—butanediisocyanate are used as
internal standards.

References — Draft CEN Method (MAFF, FScL. Norwich,
UK).

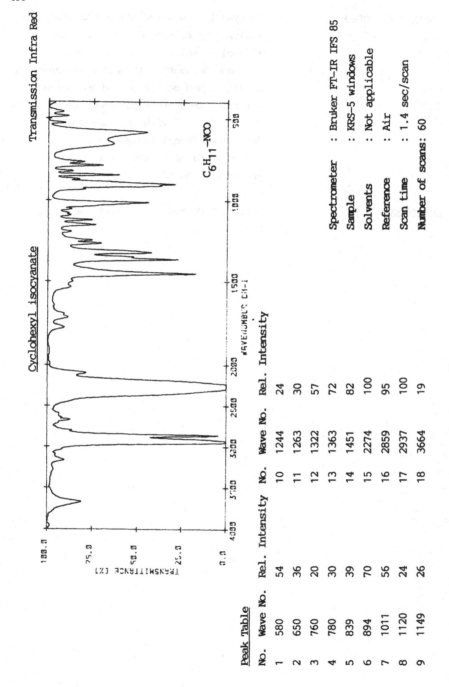

Cyclohexyl isocyanate   Transmission Infra Red

$C_6H_{11}-NCO$

Spectrometer	: Bruker FT-IR IFS 85
Sample	: KRS-5 windows
Solvents	: Not applicable
Reference	: Air
Scan time	: 1.4 sec/scan
Number of scans:	60

Peak Table

No.	Wave No.	Rel. Intensity	No.	Wave No.	Rel. Intensity
1	580	54	10	1244	24
2	650	36	11	1263	30
3	760	20	12	1322	57
4	780	30	13	1363	72
5	839	39	14	1451	82
6	894	70	15	2274	100
7	1011	56	16	2859	95
8	1120	24	17	2937	100
9	1149	26	18	3664	19

1-Decanol

$CH_3-(CH_2)_8-CH_2OH$

CAS No.	– 00112-30-1
PM Ref. No.	– 15100
Restrictions	– none
Formula	– $C_{10} H_{22} O$
Molecular weight	– 158.29
Alternative names	– n-Decyl alcohol.

Physical Characteristics – Colourless viscous liquid, mp 7^OC bp 232.9^OC. Soluble in alcohol and ether.

Handling – Store at room temperature $(25^O$C).

Safety – Irritant.

Availability – Standard sample supplied.

Current uses – Lubricant, release agent and anti-foam agent.

Applications – Mouldings and closures.

Methods of Characterisation – IR

Mass Spectroscopy

Purity – 99%

138

1-Decanol

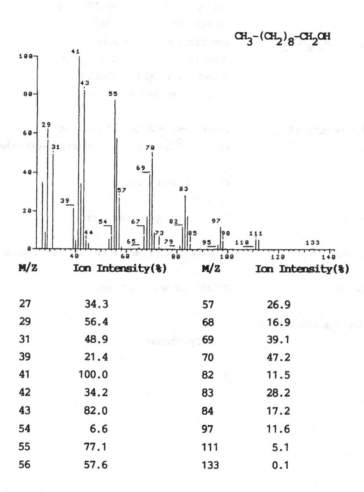

$CH_3-(CH_2)_8-CH_2OH$

M/Z	Ion Intensity(%)	M/Z	Ion Intensity(%)
27	34.3	57	26.9
29	56.4	68	16.9
31	48.9	69	39.1
39	21.4	70	47.2
41	100.0	82	11.5
42	34.2	83	28.2
43	82.0	84	17.2
54	6.6	97	11.6
55	77.1	111	5.1
56	57.6	133	0.1

Spectrometer   :Finnigan Mat SSQ 70
Inlet System   :Capillary GC/MS
Source Temperature:150°C
Electron Energy   :70 eV
Scan Range   :25-400

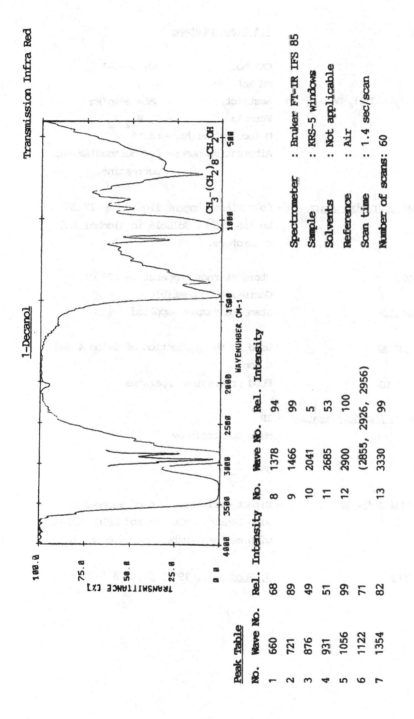

1-Decanol

Transmission Infra Red

$CH_3-(CH_2)_8-CH_2OH$

Spectrometer : Bruker FT-IR IFS 85
Sample : KRS-5 windows
Solvents : Not applicable
Reference : Air
Scan time : 1.4 sec/scan
Number of scans: 60

Peak Table

No.	Wave No.	Rel. Intensity	No.	Wave No.	Rel. Intensity
1	660	68	8	1378	94
2	721	89	9	1466	99
3	876	49	10	2041	5
4	931	51	11	2685	53
5	1056	99	12	2900	100
6	1122	71		(2855,	2926, 2956)
7	1354	82	13	3330	99

1,4-Diaminobutane

$H_2N-(CH_2)_4-NH2$

CAS No.	– 00110–60–1
PM Ref. No.	– 15250
Restrictions	– SML= 36mg/kg
Formula	– $C_4 H_{12} N_2$
Molecular weight	– 88.15
Alternative names	– 1,4–Butanediamine; putrescine.

Physical Characteristics – Colourless pungent liquid, mp 27–28°C, bp 158–160°C. Soluble in alcohol and chloroform.

Handling – Store at room temperature (25°C).

Safety – Corrosive/Combustible.

Availability – Standard sample supplied.

Current uses – Used in the production of Nylon 4 and Nylon 6.

Applications – Food processing apparatus.

Methods of Characterisation – IR
Mass Spectroscopy

Purity – 99%

Analytical methods – Direct derivatization of aqueous simulants with benzoyl chloride and HPLC analysis with methanol/water and UV detection.

References – J. Food Sci., 1991, 56, 158–160.

1,4-Diaminobutane

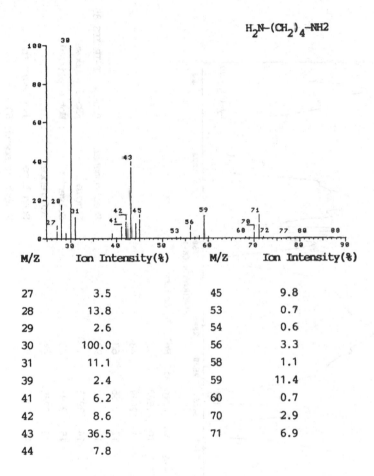

$H_2N-(CH_2)_4-NH2$

M/Z	Ion Intensity(%)	M/Z	Ion Intensity(%)
27	3.5	45	9.8
28	13.8	53	0.7
29	2.6	54	0.6
30	100.0	56	3.3
31	11.1	58	1.1
39	2.4	59	11.4
41	6.2	60	0.7
42	8.6	70	2.9
43	36.5	71	6.9
44	7.8		

Spectrometer   :Finnigan Mat SSQ 70
Inlet System   :Capillary GC/MS
Source Temperature:150°C
Electron Energy   :70 eV
Scan Range   :25-400

142

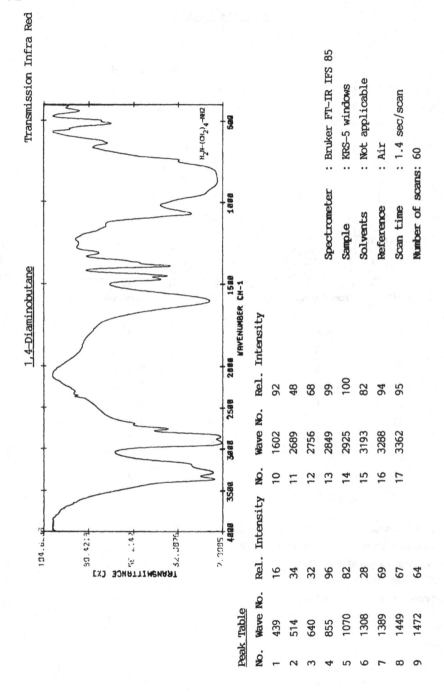

1,4-Diaminobutane

Transmission Infra Red

$H_2N-(CH_2)_4-NH_2$

Spectrometer : Bruker FT-IR IFS 85

Sample : KRS-5 windows

Solvents : Not applicable

Reference : Air

Scan time : 1.4 sec/scan

Number of scans: 60

Peak Table

No.	Wave No.	Rel. Intensity	No.	Wave No.	Rel. Intensity
1	439	16	10	1602	92
2	514	34	11	2689	48
3	640	32	12	2756	68
4	855	96	13	2849	99
5	1070	82	14	2925	100
6	1308	28	15	3193	82
7	1389	69	16	3288	94
8	1449	67	17	3362	95
9	1472	64			

Dicyclohexylmethane-4,4'-diisocyanate

CAS No.	– 05124-30-1
PM Ref. No.	– 15700
Restrictions	– QM(T)= 1mg/kg in FP (expressed as NCO).
Formula	– $C_{15} H_{22} N_2 O_2$
Molecular weight	– 262.35
Alternative names	– Desmodur W.

NCO

CH$_2$

NCO

Physical Characteristics — Colourless liquid, bp 81°C. Insoluble in water.

Handling — Store at room temperature (25°C). Protect from moisture.

Safety — Toxic.

Availability — Standard sample supplied.

Current uses — Used in the manufacture of polyurethanes.

Applications — Polyurethane tubing for food manufacturing applications. Used to make adhesives in seals for thin films, in polyester paperboard laminates (e.g susceptors) and in multi-layer high barrier plastics (e.g shelf stables) and 'boil-in-the-bag' laminates.

Methods of Characterisation — IR
Mass Spectroscopy

Purity — 99%

Analytical methods — Isocyanates in materials and articles are analysed by solvent extraction with ethanol in toluene with concurrent urethane derivative formation, clean-up by liquid/liquid partition and solid phase cartridge chromatography and determination by capillary GC with nitrogen selective detection. Phenyl isocyanate and

1,4-butanediisocyanate are used as
internal standards.

References   - Draft CEN Method (MAFF, FScL. Norwich,
UK).

Dicyclohexylmethane-4,4'-diisocyanate

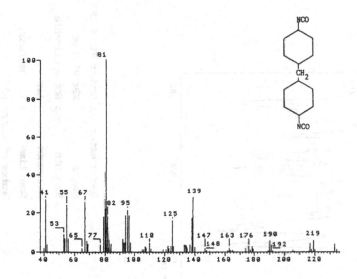

M/Z	Ion Intensity(%)	M/Z	Ion Intensity(%)
41	26.9	110	4.4
53	8.8	125	15.8
55	23.9	139	28.0
65	1.7	147	1.6
67	25.5	148	0.9
77	3.6	163	2.1
80	22.2	176	2.9
81	100.0	190	5.8
82	18.1	192	0.7
95	21.3	219	6.0

Spectrometer   : Finnigan Mat SSQ 70
Inlet System   : Capillary GC/MS
Source Temperature : 150°C
Electron Energy   : 70 eV
Scan Range   : 25–400

146

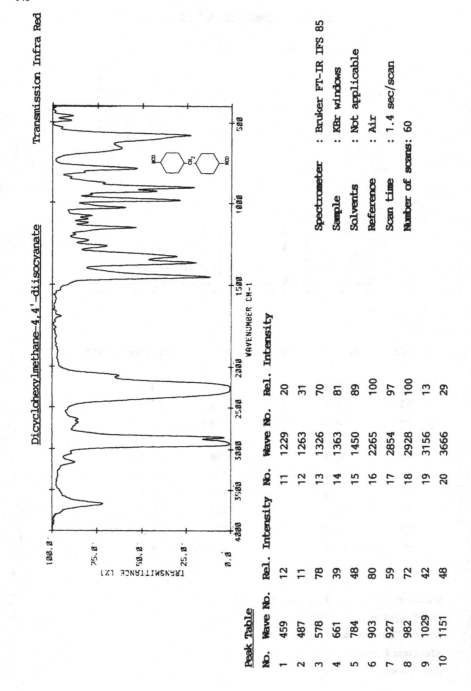

Transmission Infra Red

Dicyclohexylmethane-4,4'-diisocyanate

TRANSMITTANCE (%)

WAVENUMBER CM-1

Spectrometer	: Bruker FT-IR IFS 85
Sample	: KBr windows
Solvents	: Not applicable
Reference	: Air
Scan time	: 1.4 sec/scan
Number of scans: 60	

Peak Table

No.	Wave No.	Rel. Intensity	No.	Wave No.	Rel. Intensity
1	459	12	11	1229	20
2	487	11	12	1263	31
3	578	78	13	1326	70
4	661	39	14	1363	81
5	784	48	15	1450	89
6	903	80	16	2265	100
7	927	59	17	2854	97
8	982	72	18	2928	100
9	1029	42	19	3156	13
10	1151	48	20	3666	29

Diethyleneglycol

HO-CH$_2$-CH$_2$-O-CH$_2$-CH$_2$-OH

CAS No.	– 00111-46-6
PM Ref. No.	– 15760
Restrictions	– SML(T)=30 mg/kg
Formula	– C$_4$ H$_{10}$ O$_3$
Molecular weight	– 106.12
Alternative names	– 2,2'-Dihydroxydiethyl ether.

Physical Characteristics – Colourless liquid, mp -10.5^{o}C, bp 244.3^{o}C. Miscible with water.

Handling – Store at room temperature (25°C).

Safety – Harmful.

Availability – Standard sample supplied.

Current uses – Intermediate for polyester resins and polyethylene terephthalate (PET). Polyols for polyurethane. Esterification with rosin acids yields soft resins.

Applications – Used to make re-use and single use trays for pre-cooked, frozen and chilled meals. Carbonated beverage bottles. Roasting bags. Laminating agent and adhesive. Lacquers.

Methods of Characterisation – IR
Mass Spectroscopy

Purity – Technical grade supplied by industry. Purity not declared.

Analytical methods – Aqueous simulants concentrated then derivatized (BSTFA) to form TMS ether derivative of DEG. Determined by capillary GC with FID. Olive oil extracted with water, concentrated then treated as above.

148

References

- Draft CEN method (PIRA, Leatherhead, UK).
J. Assoc. Off. Anal. Chem., 1988, 71,
499-502.

Diethyleneglycol

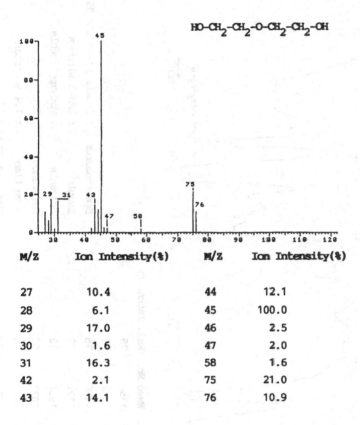

HO-CH$_2$-CH$_2$-O-CH$_2$-CH$_2$-OH

M/Z	Ion Intensity(%)	M/Z	Ion Intensity(%)
27	10.4	44	12.1
28	6.1	45	100.0
29	17.0	46	2.5
30	1.6	47	2.0
31	16.3	58	1.6
42	2.1	75	21.0
43	14.1	76	10.9

Spectrometer   :Finnigan Mat SSQ 70
Inlet System   :Capillary GC/MS
Source Temperature:150°C
Electron Energy   :70 eV
Scan Range   :25-400

150

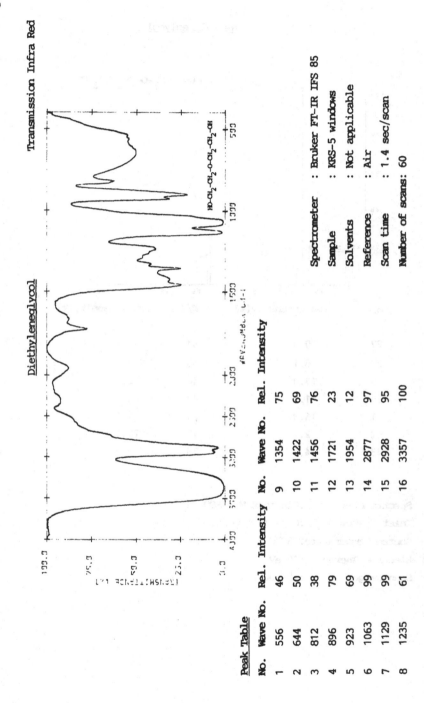

Transmission Infra Red

Diethyleneglycol

HO-CH₂-CH₂-O-CH₂-CH₂-OH

Spectrometer	: Bruker FT-IR IFS 85
Sample	: KRS-5 windows
Solvents	: Not applicable
Reference	: Air
Scan time	: 1.4 sec/scan
Number of scans: 60	

Peak Table

No.	Wave No.	Rel. Intensity	No.	Wave No.	Rel. Intensity
1	556	46	9	1354	75
2	644	50	10	1422	69
3	812	38	11	1456	76
4	896	79	12	1721	23
5	923	69	13	1954	12
6	1063	99	14	2877	97
7	1129	99	15	2928	95
8	1235	61	16	3357	100

1,2-Dihydroxybenzene

CAS No.   – 00120-80-9
PM Ref. No.   – 15880
Restrictions   – SML= 6mg/kg
Formula   – $C_6 H_6 O_2$
Molecular weight – 110.11
Alternative names– Pyrocatechol,
                        1,2-Benzenediol.

Physical Characteristics   – Pale tan powder, mp 105°C, bp 245.5°C
                                     (sublimes). Soluble in water and pyridine.

Handling   – Store at room temperature (25°C). Protect
                  from air and light.

Safety   – Corrosive.

Availability   – Standard sample supplied.

Current uses   – Stabiliser for monomers. Resin adhesives.

Applications   – Food processing machinery. Closures.

Methods of Characterisation – IR
                                          Mass Spectroscopy

Purity   – 99%

Analytical methods   – Analysis by HPLC with 55% methanol/water
                  mobile phase (ODS2 column) and UV
                  detection (277 nm). Aqueous simulants are
                  directly injected on to the HPLC whilst

olive oil is diluted with hexane,
extracted with water and injected on
to the HPLC.

References   – Method under development (MAFF, FScL,
                               Norwich, UK).

1,2-Dihydroxybenzene

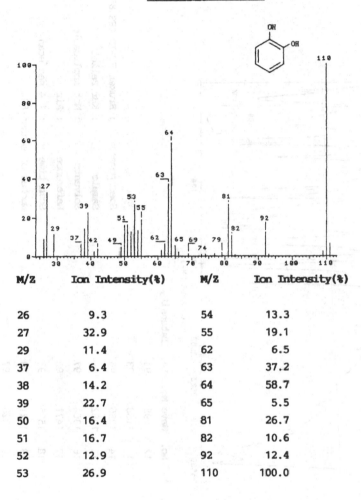

M/Z	Ion Intensity(%)	M/Z	Ion Intensity(%)
26	9.3	54	13.3
27	32.9	55	19.1
29	11.4	62	6.5
37	6.4	63	37.2
38	14.2	64	58.7
39	22.7	65	5.5
50	16.4	81	26.7
51	16.7	82	10.6
52	12.9	92	12.4
53	26.9	110	100.0

Spectrometer   :Finnigan Mat SSQ 70
Inlet System   :Capillary GC/MS
Source Temperature:150°C
Electron Energy   :70 eV
Scan Range   :25-400

Transmission Infra Red

1,2-Dihydroxybenzene

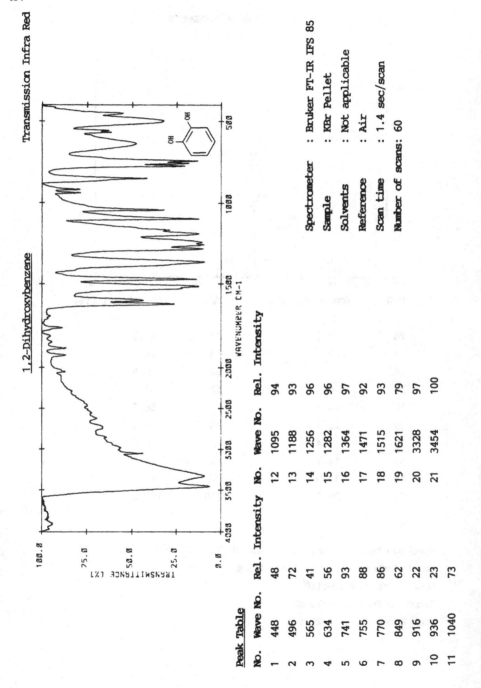

Spectrometer	: Bruker FT-IR IFS 85
Sample	: KBr Pellet
Solvents	: Not applicable
Reference	: Air
Scan time	: 1.4 sec/scan
Number of scans: 60	

Peak Table

No.	Wave No.	Rel. Intensity	No.	Wave No.	Rel. Intensity
1	448	48	12	1095	94
2	496	72	13	1188	93
3	565	41	14	1256	96
4	634	56	15	1282	96
5	741	93	16	1364	97
6	755	88	17	1471	92
7	770	86	18	1515	93
8	849	62	19	1621	79
9	916	22	20	3328	97
10	936	23	21	3454	100
11	1040	73			

1,3-Dihydroxybenzene

CAS No.	– 00108–46–3
PM Ref. No.	– 15910
Restrictions	– SML= 2.4mg/kg
Formula	– $C_6 H_6 O_2$
Molecular weight	– 110.11
Alternative names	– Resorcinol, 1,3-Benzenediol

Physical Characteristics – White crystaline powder, mp 109–111°C, bp 280°C. Soluble in ether, glycerol.

Handling – Store at room temperature (25°C). Protect from light and air.

Safety – Irritant.

Availability – Standard sample supplied.

Current uses – Reactive adjuvant employed in the production of gelatin-bonded cord compositions. Resin adhesives.

Applications – Lining for crown closures. Closures and sealing gaskets.

Methods of Characterisation – IR
Mass Spectroscopy

Purity – 99%

Analytical methods – Analysis by HPLC with 55% methanol/water mobile phase (ODS2 column) and UV

detection (277 nm). Aqueous simulants are
directly injected on to the HPLC whilst
olive oil is diluted with hexane,
extracted with water which is injected on
to the HPLC.

References   - Method under development (MAFF, FScL,
                              Norwich, UK).

1,3-Dihydroxybenzene

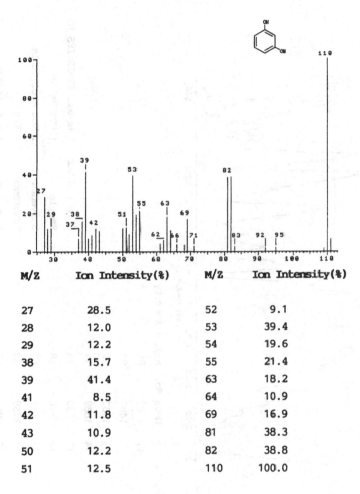

M/Z	Ion Intensity(%)	M/Z	Ion Intensity(%)
27	28.5	52	9.1
28	12.0	53	39.4
29	12.2	54	19.6
38	15.7	55	21.4
39	41.4	63	18.2
41	8.5	64	10.9
42	11.8	69	16.9
43	10.9	81	38.3
50	12.2	82	38.8
51	12.5	110	100.0

Spectrometer   :Finnigan Mat SSQ 70
Inlet System   :Capillary GC/MS
Source Temperature :150°C
Electron Energy   :70 eV
Scan Range   :25–400

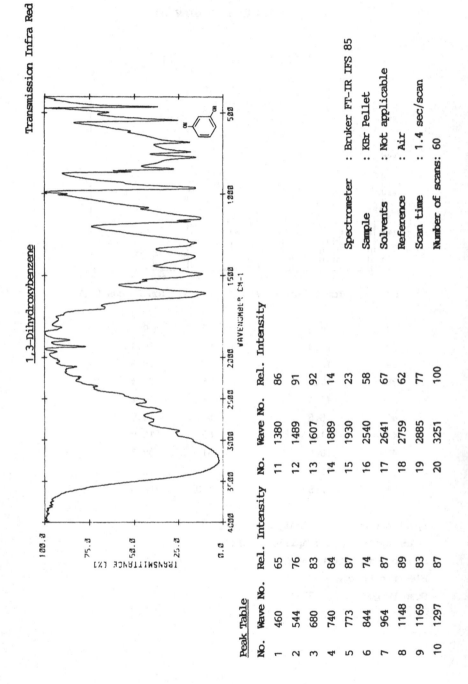

Transmission Infra Red

1,3-Dihydroxybenzene

Spectrometer	: Bruker FT-IR IFS 85
Sample	: KBr Pellet
Solvents	: Not applicable
Reference	: Air
Scan time	: 1.4 sec/scan
Number of scans: 60	

Peak Table

No.	Wave No.	Rel. Intensity	No.	Wave No.	Rel. Intensity
1	460	65	11	1380	86
2	544	76	12	1489	91
3	680	83	13	1607	92
4	740	84	14	1889	14
5	773	87	15	1930	23
6	844	74	16	2540	58
7	964	87	17	2641	67
8	1148	89	18	2759	62
9	1169	83	19	2885	77
10	1297	87	20	3251	100

1,4-Dihydroxybenzene

CAS No.	– 00123–31–9
PM Ref. No.	– 15940
Restrictions	– SML=0.6 mg/kg
Formula	– $C_6 H_6 O_2$
Molecular weight	– 110.11
Alternative names	– p-Hydroquinone,
	1,3-Benzenediol.

Physical Characteristics – White crystals, mp 171–174°C, bp 285–287°C. Soluble in alcohol, ether and water.

Handling – Store at room temperature (25°C), protect from light.

Safety – Toxic/Irritant.

Availability – Standard sample supplied.

Current uses – Manufacture of polyetherketone.

Applications – Cookware, food processing machinery and equipment, filters and conveyer belts.

Methods of Characterisation – IR
Mass Spectroscopy

Purity – 99%

Analytical methods – Analysis by HPLC with 55% methanol/water mobile phase (ODS2 column) and UV detection (277 nm). Aqueous simulants are directly injected on to the HPLC whilst olive oil is diluted with hexane, extracted with water which is injected on to the HPLC.

References – Method under development (MAFF, FScL, Norwich, UK).

1,4-Dihydroxybenzene

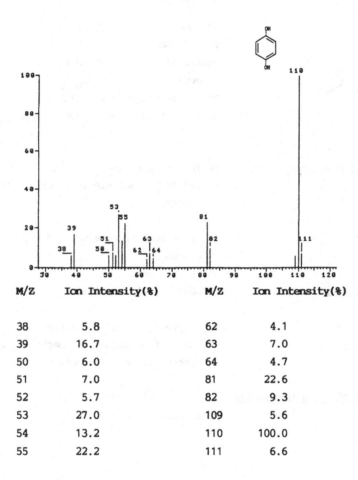

M/Z	Ion Intensity(%)	M/Z	Ion Intensity(%)
38	5.8	62	4.1
39	16.7	63	7.0
50	6.0	64	4.7
51	7.0	81	22.6
52	5.7	82	9.3
53	27.0	109	5.6
54	13.2	110	100.0
55	22.2	111	6.6

Spectrometer   :Finnigan Mat SSQ 70
Inlet System   :Capillary GC/MS
Source Temperature:150°C
Electron Energy   :70 eV
Scan Range   :25–400

Transmission Infra Red

1,4-Dihydroxybenzene

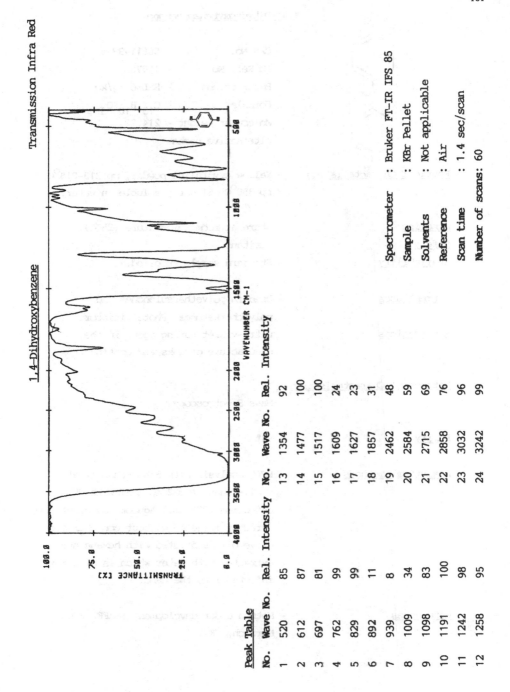

Spectrometer	: Bruker FT-IR IFS 85
Sample	: KBr Pellet
Solvents	: Not applicable
Reference	: Air
Scan time	: 1.4 sec/scan
Number of scans: 60	

Peak Table

No.	Wave No.	Rel. Intensity	No.	Wave No.	Rel. Intensity
1	520	85	13	1354	92
2	612	87	14	1477	100
3	697	81	15	1517	100
4	762	99	16	1609	24
5	829	99	17	1627	23
6	892	11	18	1857	31
7	939	8	19	2462	48
8	1009	34	20	2584	59
9	1098	83	21	2715	69
10	1191	100	22	2858	76
11	1242	98	23	3032	96
12	1258	95	24	3242	99

4,4'-Dihydroxybenzophenone

CAS No.	- 00611-99-4
PM Ref. No.	- 15970
Restrictions	- SML=6 mg/kg
Formula	- $C_{13} H_{10} O_3$
Molecular weight	- 214.22
Alternative names-	

Physical Characteristics — Yellow crystaline powder, mp 213-215°C, bp 350°C. Slightly soluble in water.

Handling — Store at room temperature (25°C).

Safety — Irritant.

Availability — Standard sample supplied.

Current uses — Used in polyethersulphones and polyetherketones. Photoinitiator.

Applications — Ultra-violet curing agent in the manufacture of inks and coatings.

Methods of Characterisation — IR
Mass Spectroscopy

Purity — 99%

Analytical methods — HPLC analysis with 55% methanol/water mobile phase (ODS2 column) with UV detection (277 nm). Aqueous simulants are directly injected without concentration. Olive oil is diluted with hexane and extracted with water which is directly injected into the HPLC.

References — Method under development (MAFF, FScL. Norwich, UK).

4,4'Dihydroxybenzophenone

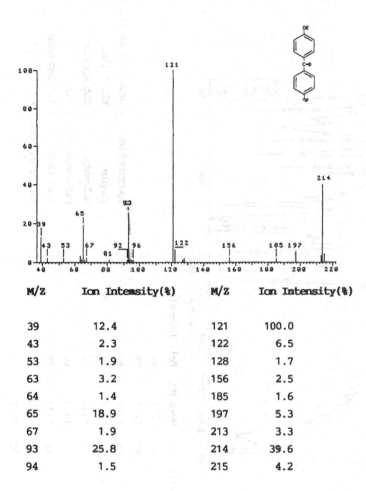

M/Z	Ion Intensity(%)	M/Z	Ion Intensity(%)
39	12.4	121	100.0
43	2.3	122	6.5
53	1.9	128	1.7
63	3.2	156	2.5
64	1.4	185	1.6
65	18.9	197	5.3
67	1.9	213	3.3
93	25.8	214	39.6
94	1.5	215	4.2

Spectrometer   :Finnigan Mat SSQ 70
Inlet System   :Capillary GC/MS
Source Temperature:150°C
Electron Energy   :70 eV
Scan Range   :25-400

164

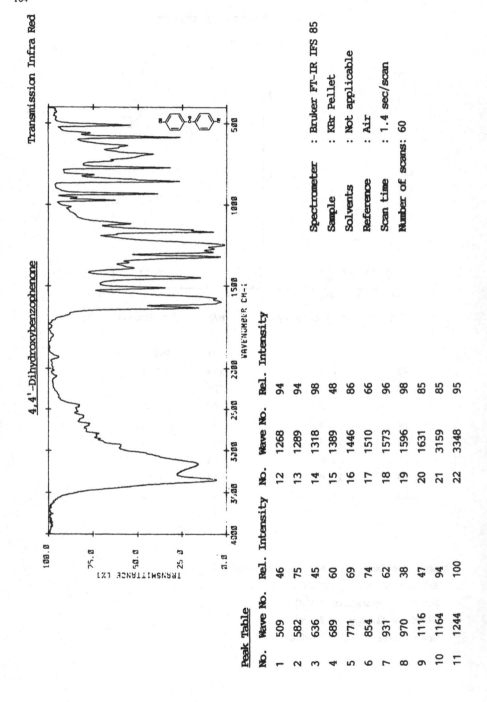

Transmission Infra Red

4,4'-Dihydroxybenzophenone

Spectrometer	: Bruker FT-IR IFS 85
Sample	: KBr Pellet
Solvents	: Not applicable
Reference	: Air
Scan time	: 1.4 sec/scan
Number of scans: 60	

Peak Table

No.	Wave No.	Rel. Intensity	No.	Wave No.	Rel. Intensity
1	509	46	12	1268	94
2	582	75	13	1289	94
3	636	45	14	1318	98
4	689	60	15	1389	48
5	771	69	16	1446	86
6	854	74	17	1510	66
7	931	62	18	1573	96
8	970	38	19	1596	98
9	1116	47	20	1631	85
10	1164	94	21	3159	85
11	1244	100	22	3348	95

4,4'-Dihydroxybiphenyl

CAS No.	– 00092-88-6
PM Ref. No.	– 16000
Restrictions	– SML=6 mg/kg
Formula	– $C_{12}H_{10}O_2$
Molecular weight	– 186.21
Alternative names	– 4,4-Biphenol.

Physical Characteristics — White powder, mp 280-282°C.

Handling — Store at room temperature (25°C).

Safety — Irritant.

Availability — Standard sample supplied.

Current uses — Used in the manufacture of polyethylene terephthalate.

Applications — Bottles for carbonated beverages.

Methods of Characterisation — IR

Mass Spectroscopy

Purity — 97%

Analytical methods — HPLC analysis with 55% methanol/water mobile phase (ODS2 column) with UV detection (277 nm). Aqueous simulants are directly injected without concentration. Olive oil is diluted with hexane and extracted with water which is directly injected into the HPLC.

References — Method under development (MAFF, FScL. Norwich, UK).

4,4'-Dihydroxybiphenyl

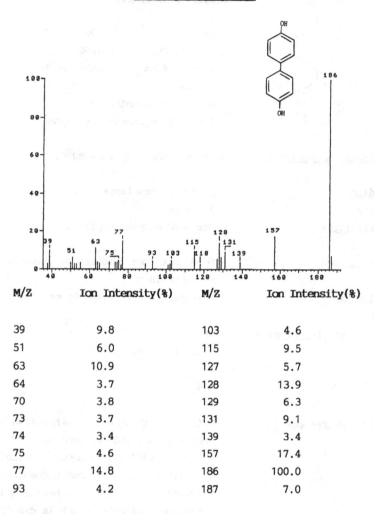

M/Z	Ion Intensity(%)	M/Z	Ion Intensity(%)
39	9.8	103	4.6
51	6.0	115	9.5
63	10.9	127	5.7
64	3.7	128	13.9
70	3.8	129	6.3
73	3.7	131	9.1
74	3.4	139	3.4
75	4.6	157	17.4
77	14.8	186	100.0
93	4.2	187	7.0

Spectrometer   :Finnigan Mat SSQ 70
Inlet System   :Capillary GC/MS
Source Temperature:150°C
Electron Energy   :70 eV
Scan Range   :25–400

Transmission Infra Red

4.4′-Dihydroxybiphenyl

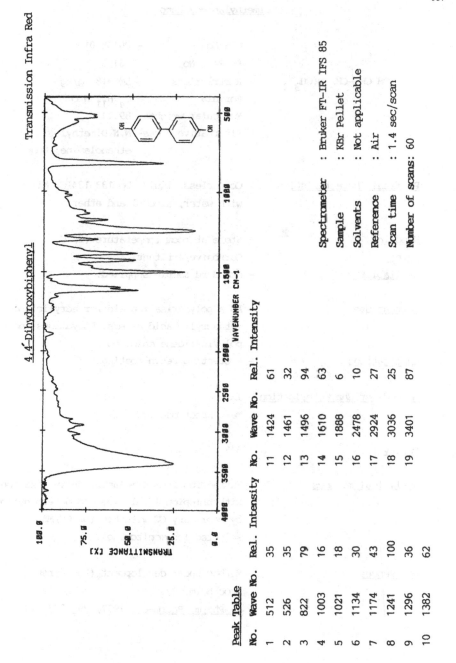

Spectrometer : Bruker FT-IR IFS 85
Sample : KBr Pellet
Solvents : Not applicable
Reference : Air
Scan time : 1.4 sec/scan
Number of scans: 60

Peak Table

No.	Wave No.	Rel. Intensity	No.	Wave No.	Rel. Intensity
1	512	35	11	1424	61
2	526	35	12	1461	32
3	822	79	13	1496	94
4	1003	16	14	1610	63
5	1021	18	15	1888	6
6	1134	30	16	2478	10
7	1174	43	17	2924	27
8	1241	100	18	3036	25
9	1296	36	19	3401	87
10	1382	62			

Dimethylaminoethanol

$OH-CH_2-CH_2-N(CH_3)_2$

CAS No.	– 00108–01–0
PM Ref. No.	– 16150
Restrictions	– SML=18 mg/kg
Formula	– $C_4 H_{11} N O$
Molecular weight	– 89.14
Alternative names–	N,N–Dimethyl-ethanolamine; Deanol.

Physical Characteristics – Colourless liquid, bp 133–134°C. Miscible with water, alcohol and ether.

Handling – Store at room temperature (25°C).

Safety – Corrosive/Irritant.

Availability – Standard sample supplied.

Current uses – As a polymerisation aid for acrylic and methacrylic acid esters. Polyamides and polyvinylidene chloride.

Applications – Protective resin coatings.

Methods of Characterisation – IR

Mass Spectroscopy

Purity – 64%

Analytical methods – Extraction from simulants, derivatization with priopionyl chloride and determination by capillary GC with FID (or GC/MS selected ion monitoring).

References – Method under development (Packforsk, Stockholm, S).

J. Pharm. Pharmac., 1977, 29, 373–374.

Dimethylaminoethanol

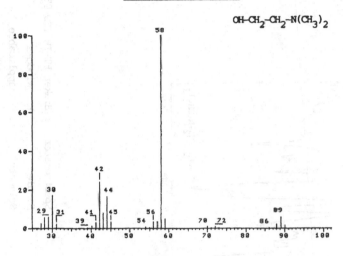

$OH-CH_2-CH_2-N(CH_3)_2$

M/Z	Ion Intensity(%)	M/Z	Ion Intensity(%)
27	2.4	45	3.3
28	5.6	56	3.3
29	5.8	57	3.3
30	17.1	58	100.0
31	2.0	59	4.8
40	1.2	70	0.8
41	3.0	72	0.8
42	24.1	88	2.0
43	8.0	89	6.0
44	16.7	90	1.4

Spectrometer   :Finnigan Mat SSQ 70
Inlet System   :Capillary GC/MS
Source Temperature:150°C
Electron Energy   :70 eV
Scan Range   :25-400

170

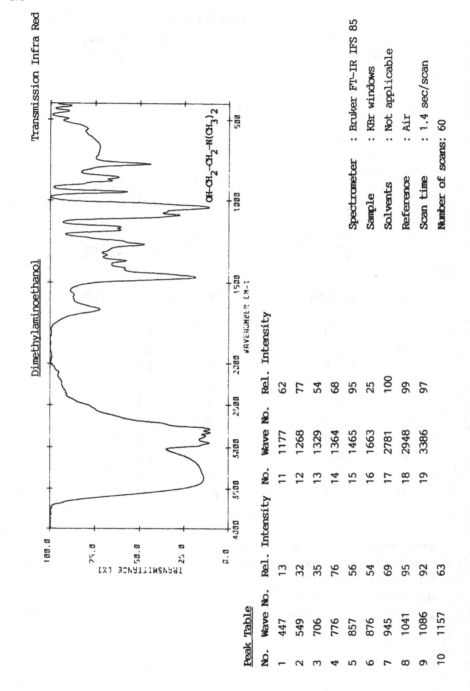

Transmission Infra Red

Dimethylaminoethanol

OH-CH₂-CH₂-N(CH₃)₂

Spectrometer	: Bruker FT-IR IFS 85
Sample	: KBr windows
Solvents	: Not applicable
Reference	: Air
Scan time	: 1.4 sec/scan
Number of scans: 60	

Peak Table

No.	Wave No.	Rel. Intensity	No.	Wave No.	Rel. Intensity
1	447	13	11	1177	62
2	549	32	12	1268	77
3	706	35	13	1329	54
4	776	76	14	1364	68
5	857	56	15	1465	95
6	876	54	16	1663	25
7	945	69	17	2781	100
8	1041	95	18	2948	99
9	1086	92	19	3386	97
10	1157	63			

3,3'-Dimethyl-4,4'-diisocyanatobiphenyl

CAS No.	- 00091-97-4
PM Ref. No.	- 16240
Restrictions	- QM(T)= 1mg/kg
	(expressed as NCO)
Formula	- $C_{16} H_{12} N_2 O_2$
Molecular weight	- 264.29
Alternative names-	

Physical Characteristics -

Safety -
Availability - No sample supplied.

Current uses - Used in the manufacture of polyurethanes.
Applications - Polyurethane tubing for food manufacturing applications. Used to make adhesives in seals for thin films, in polyester paperboard laminates (e.g susceptors) and in multi-layer high barrier plastics (e.g shelf stables) and in 'boil-in-the-bag' laminates.

Analytical methods -
References - See other isocyanate entries.

Dipentaerythritol

$(HOCH_2)_3C-CH_2-O-CH_2-C(CH_2OH)_3$

CAS No.	– 0126–58–9
PM Ref. No.	– 16480
Restrictions	– none
Formula	– $C_{10} H_{22} O_7$
Molecular weight	– 254.28
Alternative names–	

Physical Characteristics — White crystalline powder, mp 221°C. Slightly soluble in water.

Handling — Store at room temperature (25°C).

Safety — Flammable.

Availability — Standard sample supplied.

Current uses — Used to cross-link unsaturated polyester resins. Modifying rosin. Imparts stability to PVC latexes.

Applications — Adhesives. Coatings. Lubricants. Films.

Methods of Characterisation — IR
Mass Spectroscopy

Purity — 99%

Dipentaerythritol

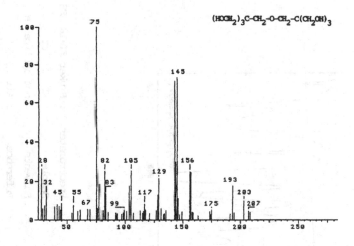

$(HOCH_2)_3C-CH_2-O-CH_2-C(CH_2OH)_3$

M/Z	Ion Intensity(%)	M/Z	Ion Intensity(%)
28	26.6	117	8.2
32	13.3	129	21.6
45	8.7	143	71.0
55	7.8	144	29.9
67	5.5	145	72.9
75	100.0	156	24.9
82	25.2	175	5.1
83	13.3	193	17.5
99	5.1	203	9.9
105	25.4	207	4.5

Spectrometer   :Finnigan Mat SSQ 70
Inlet System   :Capillary GC/MS
Source Temperature :150°C
Electron Energy   :70 eV
Scan Range   :25–400

Dipentaerythritol   Transmission Infra Red

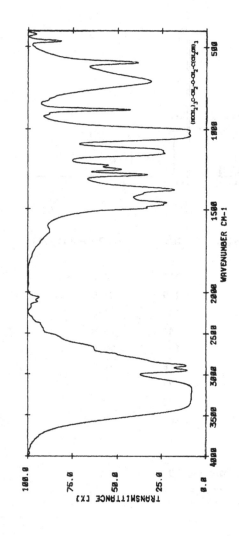

(HOCH$_2$)$_3$C-CH$_2$-O-CH$_2$-C(CH$_2$OH)$_3$

Spectrometer	: Bruker FT-IR IFS 85
Sample	: KBr Pellet
Solvents	: Not applicable
Reference	: Air
Scan time	: 1.4 sec/scan
Number of scans: 60	

Peak Table

No.	Wave No.	Rel. Intensity	No.	Wave No.	Rel. Intensity
1	464	17	8	1249	44
2	594	52	9	1281	58
3	709	62	10	1376	78
4	878	48	11	1456	72
5	1035	92	12	2888	91
6	1149	69	13	2943	92
7	1224	38	14	3259	100

175

Diphenylether-4,4'-diisocyanate

CAS No.	– 04128-73-8
PM Ref. No.	– 16570
Restrictions	– QM(T)= 1mg/kg
Formula	– $C_{14} H_8 N_2 O_3$
Molecular weight	– 252.23
Alternative names-	

Physical Characteristics –

Safety –

Availability – No sample supplied.

Current uses – Used in the manufacture of polyurethanes.

Applications – Polyurethane tubing for food manufacturing applications. Used to make adhesives in seals for thin films, in polyester paperboard laminates (e.g susceptors) and in multi-layer high barrier plastics (e.g shelf stables) and 'boil-in-the-bag' laminates.

Analytical methods – Isocyanates in materials and articles are analysed by solvent extraction with ethanol in toluene with concurrent urethane derivative formation, clean-up by liquid/liquid partition and solid phase cartridge chromatography and determination by capillary GC with nitrogen selective detection. Phenyl isocyanate and 1,4-butanediisocyanate are used as internal standards.

References – Draft CEN Method (MAFF, FScL. Norwich, UK).

Diphenylmethane-2,4'-diisocyanate

CAS No.	– 05873-54-1
PM Ref. No.	– 16600
Restrictions	– QM(T)= 1mg/kg in FP (expressed as NCO).
Formula	– $C_{15} H_{10} N_2 O_2$
Molecular weight	– 250.16
Alternative names	– Desmodur CD.

Physical Characteristics
– Pale yellow liquid, bp $230^{o}C$. Soluble in most organic solvents.

Handling
– Room temperature ($25^{o}C$). Protect from moisture.

Safety
– Harmful/Irritant.

Availability
– Standard sample supplied.

Current uses
– Used to make flexible polyurethane foams and other urethane polymers.

Applications
– Polyurethane tubing for food manufacturing applications. Used to make adhesives in seals for thin films, in polyester paperboard laminates (e.g susceptors) and in multi-layer high barrier plastics (e.g shelf stables) and 'boil-in-the-bag' laminates. Baking enamels.

Methods of Characterisation
– IR
Mass Spectroscopy

Purity
– 90% (contains the 2,6' isomer).

Analytical methods
– Isocyanates in materials and articles are analysed by solvent extraction with ethanol in toluene with concurrent

urethane derivative formation, clean-up by liquid/liquid partition and solid phase cartridge chromatography and determination by capillary GC with nitrogen selective detection. Phenyl isocyanate and 1,4-butanediisocyanate are used as internal standards.

References   - Draft CEN Method (MAFF, FScL. Norwich, UK).

Diphenylmethane-2,4'-diisocyanate

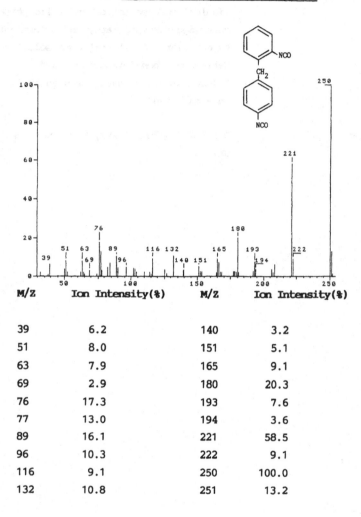

M/Z	Ion Intensity(%)	M/Z	Ion Intensity(%)
39	6.2	140	3.2
51	8.0	151	5.1
63	7.9	165	9.1
69	2.9	180	20.3
76	17.3	193	7.6
77	13.0	194	3.6
89	16.1	221	58.5
96	10.3	222	9.1
116	9.1	250	100.0
132	10.8	251	13.2

Spectrometer   :Finnigan Mat SSQ 70
Inlet System   :Capillary GC/MS
Source Temperature:150°C
Electron Energy   :70 eV
Scan Range   :25-400

Transmission Infra Red

Diphenylmethane-2,4'-diisocyanate

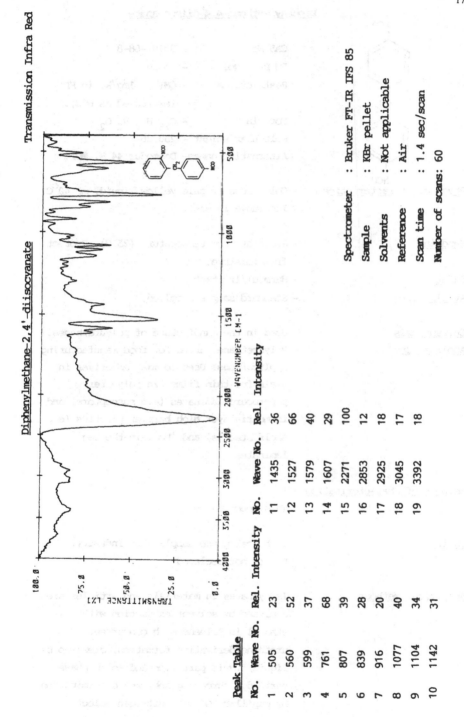

Spectrometer	: Bruker FT-IR IFS 85
Sample	: KBr pellet
Solvents	: Not applicable
Reference	: Air
Scan time	: 1.4 sec/scan
Number of scans: 60	

Peak Table

No.	Wave No.	Rel. Intensity	No.	Wave No.	Rel. Intensity
1	505	23	11	1435	36
2	560	52	12	1527	66
3	599	37	13	1579	40
4	761	68	14	1607	29
5	807	39	15	2271	100
6	839	28	16	2853	12
7	916	20	17	2925	18
8	1077	40	18	3045	17
9	1104	34	19	3392	18
10	1142	31			

Diphenylmethane-4,4'-diisocyanate

NCO

CH$_2$

NCO

CAS No.	– 00101-68-8
PM Ref. No.	– 16630
Restrictions	– QM(T)= 1mg/kg in FP
	(expressed as NCO).
Formula	– C$_{15}$ H$_{10}$ N$_2$ O$_2$
Molecular weight	– 250.26
Alternative names	– Desmodur 44 M, MDI.

Physical Characteristics — Colourless or pale yellow liquid, mp 40°C. Insoluble in water.

Handling — Store at room temperature (25°C). Protect from moisture.

Safety — Harmful/Irritant.

Availability — Standard sample supplied.

Current uses — Used in the manufacture of polyurethanes.

Applications — Polyurethane tubing for food manufacturing applications. Used to make adhesives in seals for thin films, in polyester paperboard laminates (e.g susceptors) and in multi-layer high barrier plastics (e.g shelf stables) and 'boil-in-the-bag' laminates.

Methods of Characterisation — IR
Mass Spectroscopy

Purity — Technical grade supplied by industry. Purity not declared.

Analytical methods — Isocyanates in materials and articles are analysed by solvent extraction with ethanol in toluene with concurrent urethane derivative formation, clean-up by liquid/liquid partition and solid phase cartridge chromatography and determination by capillary GC with nitrogen selective

detection. Phenyl isocyanate and
1,4-butanediisocyanate are used as
internal standards.

References   - Draft CEN Method (MAFF, FScL. Norwich,
UK).

Diphenylmethane-4,4'-diisocyanate

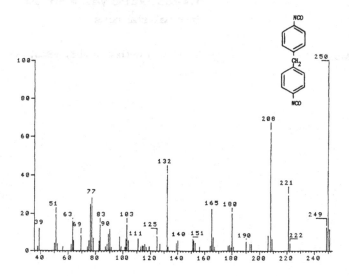

M/Z	Ion Intensity(%)	M/Z	Ion Intensity(%)
39	11.8	132	40.1
51	19.2	140	5.5
63	16.0	151	6.5
69	8.3	165	22.4
77	28.3	180	20.0
83	14.0	190	5.0
90	11.5	208	62.6
103	14.0	221	29.6
111	6.3	249	12.6
125	8.0	250	100.0

Spectrometer   :Finnigan Mat SSQ 70
Inlet System   :Capillary GC/MS
Source Temperature:150°C
Electron Energy   :70 eV
Scan Range   :25-400

183

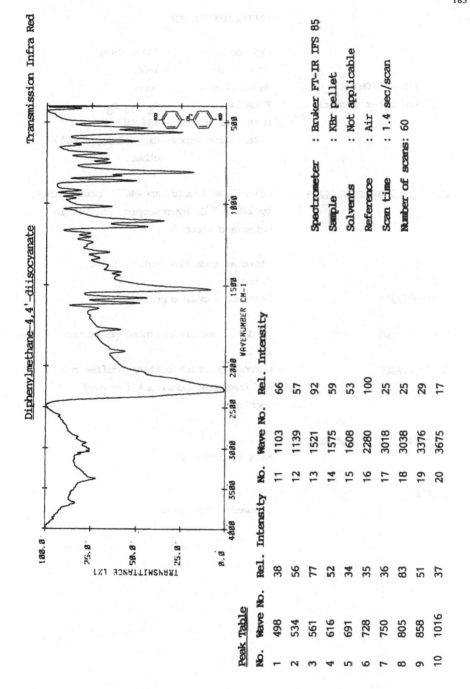

Transmission Infra Red

Diphenylmethane-4,4'-diisocyanate

Spectrometer	: Bruker FT-IR IFS 85
Sample	: KBr pellet
Solvents	: Not applicable
Reference	: Air
Scan time	: 1.4 sec/scan
Number of scans: 60	

Peak Table

No.	Wave No.	Rel. Intensity	No.	Wave No.	Rel. Intensity
1	498	38	11	1103	66
2	534	56	12	1139	57
3	561	77	13	1521	92
4	616	52	14	1575	59
5	691	34	15	1608	53
6	728	35	16	2280	100
7	750	36	17	3018	25
8	805	83	18	3038	25
9	858	51	19	3376	29
10	1016	37	20	3675	17

Dipropyleneglycol

$[CH_3-CH(OH)CH_2]_2O$
and other isomers.

CAS No.	– 00110-98-5
PM Ref. No.	– 16660
Restrictions	– none
Formula	– $C_6 H_{14} O_3$
Molecular weight	– 134.18
Alternative names	– Bis(hydroxypropyl) ether.

Physical Characteristics — Colourless liquid, mp -40^OC (pour point), bp 229-232°C. Hygroscopic. Soluble in water and alcohol.

Handling — Store at room temperature (25°C).
Safety — Irritant.
Availability — Standard sample supplied.

Current uses — Used to make cross-linked polyester resins.
Applications — Carbonated drink bottles, coffee makers and toasters. Lubricant for food machinery.

Methods of Characterisation — IR
Mass Spectroscopy

Purity — 99%
(Isomeric mixture).

Dipropyleneglycol (isomer)

$[CH_3-CH(OH)CH_2]_2O$
and other isomers.

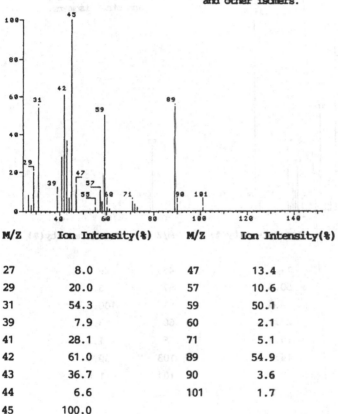

M/Z	Ion Intensity(%)	M/Z	Ion Intensity(%)
27	8.0	47	13.4
29	20.0	57	10.6
31	54.3	59	50.1
39	7.9	60	2.1
41	28.1	71	5.1
42	61.0	89	54.9
43	36.7	90	3.6
44	6.6	101	1.7
45	100.0		

Spectrometer   :Finnigan Mat SSQ 70
Inlet System   :Capillary GC/MS
Source Temperature: 150°C
Electron Energy   :70 eV
Scan Range   :25–400

Dipropyleneglycol (isomer)

$[CH_3-CH(OH)CH_2]_2O$
and other isomers.

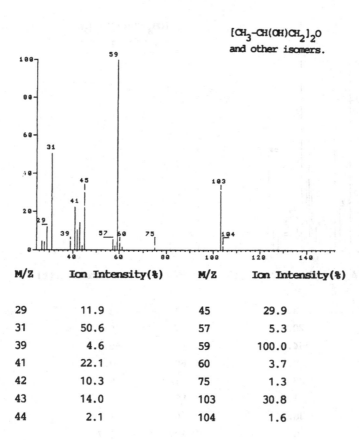

M/Z	Ion Intensity(%)	M/Z	Ion Intensity(%)
29	11.9	45	29.9
31	50.6	57	5.3
39	4.6	59	100.0
41	22.1	60	3.7
42	10.3	75	1.3
43	14.0	103	30.8
44	2.1	104	1.6

Spectrometer   :Finnigan Mat SSQ 70
Inlet System   :Capillary GC/MS
Source Temperature:150°C
Electron Energy   :70 eV
Scan Range   :25-400

Dipropyleneglycol (isomer)

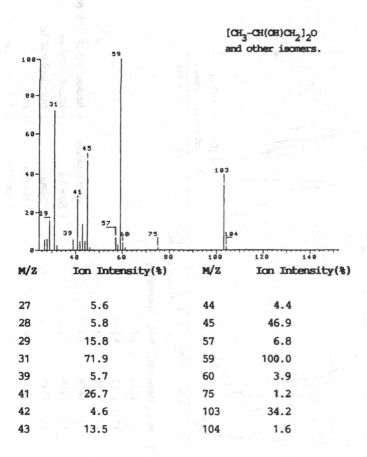

$[CH_3-CH(OH)CH_2]_2O$
and other isomers.

M/Z	Ion Intensity(%)	M/Z	Ion Intensity(%)
27	5.6	44	4.4
28	5.8	45	46.9
29	15.8	57	6.8
31	71.9	59	100.0
39	5.7	60	3.9
41	26.7	75	1.2
42	4.6	103	34.2
43	13.5	104	1.6

Spectrometer   :Finnigan Mat SSQ 70
Inlet System   :Capillary GC/MS
Source Temperature:150°C
Electron Energy   :70 eV
Scan Range   :25-400

Dipropyleneglycol

Transmission Infra Red

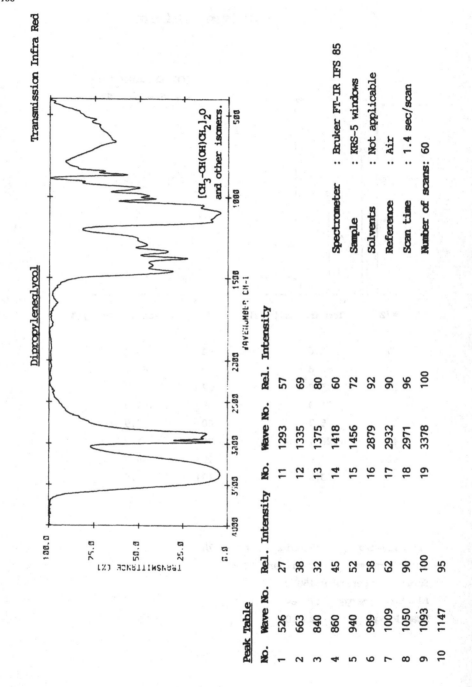

$[CH_3-CH(OH)CH_2]_2O$ and other isomers.

Spectrometer : Bruker FT-IR IFS 85
Sample : KRS-5 windows
Solvents : Not applicable
Reference : Air
Scan time : 1.4 sec/scan
Number of scans: 60

Peak Table

No.	Wave No.	Rel. Intensity	No.	Wave No.	Rel. Intensity
1	526	27	11	1293	57
2	663	38	12	1335	69
3	840	32	13	1375	80
4	860	45	14	1418	60
5	940	52	15	1456	72
6	989	58	16	2879	92
7	1009	62	17	2932	90
8	1050	90	18	2971	96
9	1093	100	19	3378	100
10	1147	95			

Epichlorohydrin

$$Cl-CH_2-\overset{\displaystyle O}{\overset{\diagup\diagdown}{CH-CH_2}}$$

CAS No.	– 00106–89–8
PM Ref. No.	– 16750
Restrictions	– Qm= 1mg/kg in FP
Formula	– $C_3 H_5 Cl O$
Molecular weight	– 92.53
Alternative names	– 1–Chloro–2,3–epoxypropane.

Physical Characteristics – Colourless liquid, mp -57°C, bp 117.9°C.

Handling – Store at room temperature (25°C).

Safety – Carcinogen/Toxic/Corrosive.

Availability – Standard sample supplied.

Current uses – Epoxy resins with p-hydroxybenzoic acid. Resins with dimethylamine. Starting substance reacted with Bisphenol A to give Bisphenol A diglycidyl ether (BADGE) used in epoxy resins, which are cross-linked to hardeners.

Applications – Coatings for cans for fruit, vegetables, & beverages. Also for coating storage vats and silos for wine, beer, fats and dry foods. Adhesives.

Methods of Characterisation – IR
Mass Spectroscopy

Purity – 99%

190

| Analytical methods | – Solvent extraction (diethyl ether) from polymer or coating and GC analysis or direct headspace analysis of material using 'hot jar' technique. Detection by electron capture. |

References

– Method under development (TNO, Zeist, NL).
J. Chromatogr., 1988, 439, 448–452.
J. Chromatogr., 1981, 398, 398–402.

Epichlorohydrin

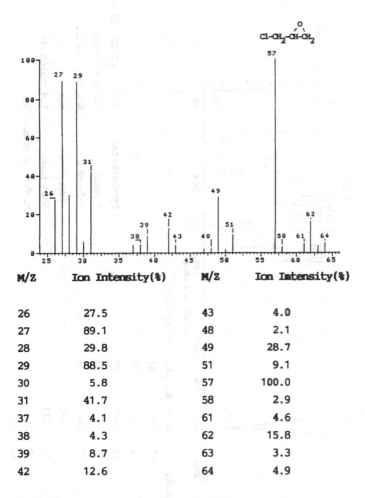

M/Z	Ion Intensity(%)	M/Z	Ion Intensity(%)
26	27.5	43	4.0
27	89.1	48	2.1
28	29.8	49	28.7
29	88.5	51	9.1
30	5.8	57	100.0
31	41.7	58	2.9
37	4.1	61	4.6
38	4.3	62	15.8
39	8.7	63	3.3
42	12.6	64	4.9

Spectrometer   :Finnigan Mat SSQ 70
Inlet System   :Capillary GC/MS
Source Temperature:150°C
Electron Energy   :70 eV
Scan Range   :25-400

Epichlorohydrin

Transmission Infra Red

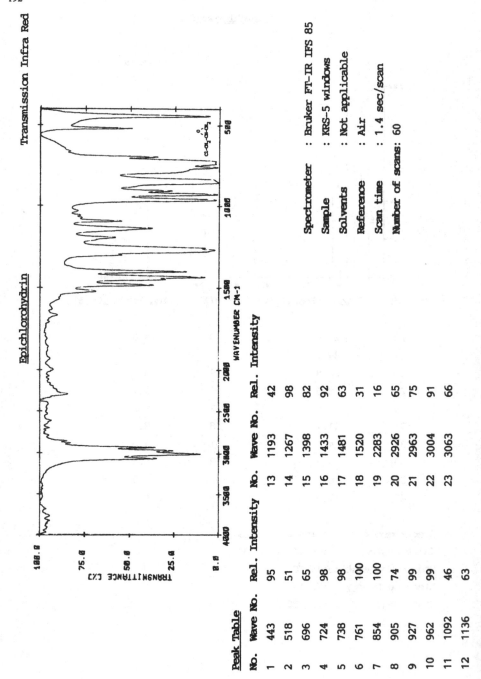

Spectrometer	: Bruker FT-IR IFS 85
Sample	: KRS-5 windows
Solvents	: Not applicable
Reference	: Air
Scan time	: 1.4 sec/scan
Number of scans: 60	

Peak Table

No.	Wave No.	Rel. Intensity	No.	Wave No.	Rel. Intensity
1	443	95	13	1193	42
2	518	51	14	1267	98
3	696	65	15	1398	82
4	724	98	16	1433	92
5	738	98	17	1481	63
6	761	100	18	1520	31
7	854	100	19	2283	16
8	905	74	20	2926	65
9	927	99	21	2963	75
10	962	99	22	3004	91
11	1092	46	23	3063	66
12	1136	63			

Ethanol

CH_3-CH_2-OH

CAS No.	— 00064-17-5
PM Ref. No.	— 16780
Restrictions	— none
Formula	— $C_2 H_6 OH$
Molecular weight	— 46.07
Alternative names	— Ethyl alcohol; absolute alcohol.

Physical Characteristics — Colourless liquid, mp -114.1°C, bp 78.5°C. Miscible with water and many organic liquids.

Handling — Store at room temperature (25°C).

Safety — Flammable.

Availability — Standard sample supplied.

Current uses — Used as a chemical intermediate and solvent. A starting substance in the synthesis of ethyl acetate and ethyl acrylate and ethyl vinyl ether.

Applications — Lacquers, coatings and adhesives. Films.

Methods of Characterisation — IR
Mass Spectroscopy

Purity — 99.7%

Ethanol

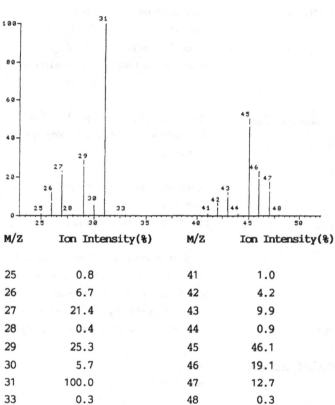

M/Z	Ion Intensity(%)	M/Z	Ion Intensity(%)
25	0.8	41	1.0
26	6.7	42	4.2
27	21.4	43	9.9
28	0.4	44	0.9
29	25.3	45	46.1
30	5.7	46	19.1
31	100.0	47	12.7
33	0.3	48	0.3

Spectrometer   :Finnigan Mat SSQ 70
Inlet System   :Capillary GC/MS
Source Temperature:150°C
Electron Energy   :70 eV
Scan Range   :25-400

Ethylene

$CH_2{=}CH_2$

CAS No.	— 00074–85–1
PM Ref. No.	— 16950
Restrictions	— none
Formula	— C_2H_4
Molecular weight	— 28.05
Alternative names	— Ethene.

Physical Characteristics — Colourless gas, mp −169.4°C, bp −103.7°C. Sparingly soluble in water. Soluble in alcohol, acetone and benzene.

Safety — Flammable.

Availability — No sample supplied.

Current uses — Used to make thermoplastics. Polyethylene film (LLDPE, LDPE, HDPE). Polyester resins. Used to make acrylate ester co-polymers. Co-polymer with vinyl acetate and vinyl alcohol (EVA, EVOH).

Applications — LDPE-films for pre-packed fresh and frozen foods, bakery products, dairy foods and desserts. Part of laminate with paper or aluminium. Coating on cartons for frozen foods. Made into bags, & blow moulded for storage containers. HDPE-bottles for milk, beakers, bags, tubs, crates, trays, and utensils. EVA-laminate for paperboard for ice-cream. EVOH-Bottles for ketchup and mayonaise. Shaped squeezable containers and closures. Barrier films.

196

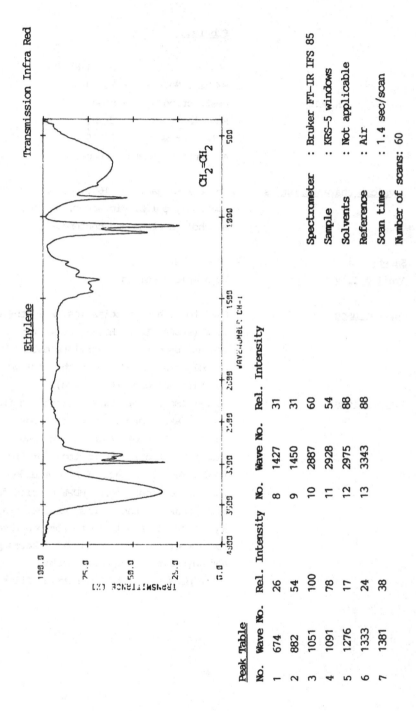

Ethylene   Transmission Infra Red

$CH_2=CH_2$

Spectrometer : Bruker FT-IR IFS 85
Sample   : KRS-5 windows
Solvents   : Not applicable
Reference   : Air
Scan time   : 1.4 sec/scan
Number of scans: 60

Peak Table

No.	Wave No.	Rel. Intensity	No.	Wave No.	Rel. Intensity
1	674	26	8	1427	31
2	882	54	9	1450	31
3	1051	100	10	2887	60
4	1091	78	11	2928	54
5	1276	17	12	2975	88
6	1333	24	13	3343	88
7	1381	38			

Ethylenediamine

$$NH_2-CH_2-CH_2-NH_2$$

CAS No.	– 00107–15–3
PM Ref. No.	– 16960
Restrictions	– SML= 12mg/kg
Formula	– $C_2\,H_8\,N_2$
Molecular weight	– 60.10
Alternative names	– 1,2-Ethanediamine.

Physical Characteristics — Colourless viscous liquid, mp 8.5°C, bp 118°C. Soluble in water and alcohol.

Handling — Store at room temperature (25°C). Protect from air.

Safety — Toxic/Corrosive.

Availability — Standard sample supplied.

Current uses — Used to make some nylons, and thermosetting resins. Reactive hardeners in epoxy resins. Stabilising rubber latexes.

Applications — Moisture barrier coatings for paper, cellophane etc. Adhesives. Corrosion inhibitor for aluminiium alloys.

Methods of Characterisation — IR
Mass Spectroscopy

Purity — 99%

Analytical methods —

References — Method under development (Fraunhofer Inst., Munich, D).

Ethylenediamine

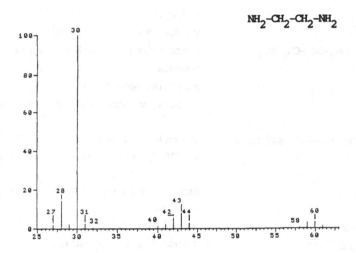

$NH_2-CH_2-CH_2-NH_2$

M/Z	Ion Intensity(%)	M/Z	Ion Intensity(%)
27	1.7	41	2.2
28	14.0	42	5.1
29	2.2	43	7.7
30	100.0	44	2.7
31	2.4	58	0.2
32	0.4	59	2.6
39	0.3	60	3.0
40	1.0	61	1.1

Spectrometer   :Finnigan Mat SSQ 70
Inlet System   :Capillary GC/MS
Source Temperature :150°C
Electron Energy   :70 eV
Scan Range   :25–400

199

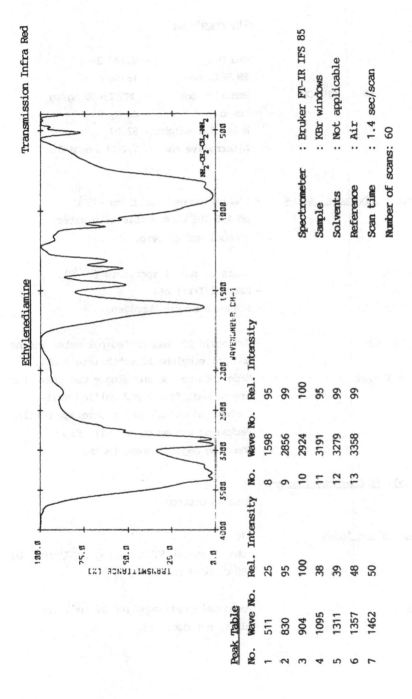

Transmission Infra Red

Ethylenediamine

Spectrometer : Bruker FT-IR IFS 85
Sample : KBr windows
Solvents : Not applicable
Reference : Air
Scan time : 1.4 sec/scan
Number of scans: 60

Peak Table

No.	Wave No.	Rel. Intensity	No.	Wave No.	Rel. Intensity
1	511	25	8	1598	95
2	830	95	9	2856	99
3	904	100	10	2924	100
4	1095	38	11	3191	95
5	1311	39	12	3279	99
6	1357	48	13	3358	99
7	1462	50			

Ethyleneglycol

HO-CH$_2$-CH$_2$-OH

CAS No.   — 00107-21-1
PM Ref. No.   — 16990
Restrictions   — SML(T)=30 mg/kg
Formula   — C$_2$ H$_6$ O$_2$
Molecular weight   — 62.07
Alternative names— 1,2-Ethanediol.

Physical Characteristics — Clear viscous liquid, mp -13°C,
bp 196-198°C. Miscible with water,
alcohol and glycerol.

Handling   — Store at room temperature (25°C).
Safety   — Harmful/Irritant.
Availability — Standard sample supplied.

Current uses — Synthesis of unsaturated polyester resins
and polyethylene terephthalate (PET).

Applications — Made into re-use and single use trays for
pre-cooked, frozen and chilled meals.
Carbonated and alcoholic beverage bottles.
Packaging for meats and oil. Films.
Roasting bags. Storage tanks.

Methods of Characterisation — IR
Mass Spectroscopy

GC Retention Index — 376
(DB5, 3 min at 50°C, rising 20°C/min^{-1} to
300°C, hold for 20 min.)

Purity — Technical grade supplied by industry.
Purity not declared.

Analytical methods   – Capillary GC (FID) – concentration of
aqueous simulants by evaporation then
derivatization (TMS ethers). Olive oil
extracted with water, concentrated and
derivatized for GC.

References   – Draft CEN method (PIRA, UK).
J. Assoc. Off. Anal. Chem., 1988, 71,
499–502
Food Additives and Contaminants, 1988, 5,
485–492

Ethyleneglycol

$HO-CH_2-CH_2-OH$

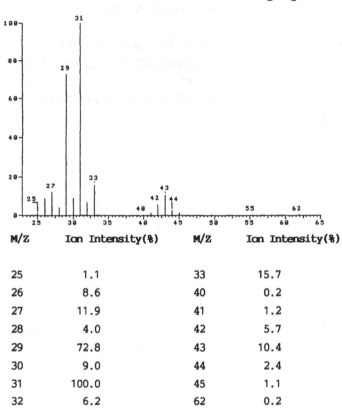

M/Z	Ion Intensity(%)	M/Z	Ion Intensity(%)
25	1.1	33	15.7
26	8.6	40	0.2
27	11.9	41	1.2
28	4.0	42	5.7
29	72.8	43	10.4
30	9.0	44	2.4
31	100.0	45	1.1
32	6.2	62	0.2

Spectrometer   :Finnigan Mat SSQ 70
Inlet System   :Capillary GC/MS
Source Temperature:150°C
Electron Energy   :70 eV
Scan Range   :25–400

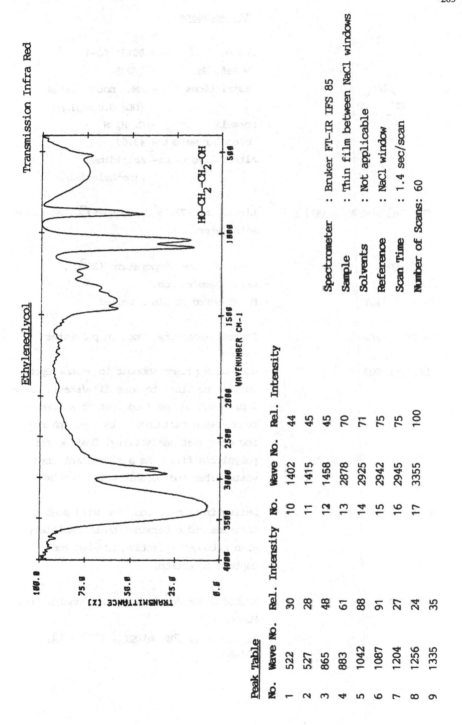

Ethyleneglycol

Transmission Infra Red

HO-CH$_2$-CH$_2$-OH

WAVENUMBER CM-1

TRANSMITTANCE [%]

Spectrometer	: Bruker FT-IR IFS 85
Sample	: Thin film between NaCl windows
Solvents	: Not applicable
Reference	: NaCl window
Scan Time	: 1.4 sec/scan
Number of Scans: 60	

Peak Table

No.	Wave No.	Rel. Intensity	No.	Wave No.	Rel. Intensity
1	522	30	10	1402	44
2	527	28	11	1415	45
3	865	48	12	1458	45
4	883	61	13	2878	70
5	1042	88	14	2925	71
6	1087	91	15	2942	75
7	1204	27	16	2945	75
8	1256	24	17	3355	100
9	1335	35			

Ethyleneimine

CH$_2$ - CH$_2$ with NH bridge

CAS No.	- 00151-56-4
PM Ref. No.	- 17005
Restrictions	- SML= not detectable (DL= 0.01mg/kg).
Formula	- C$_2$ H$_5$ N
Molecular weight	- 43.07
Alternative names	- Aziridine, Dimethylenimine.

Physical Characteristics - Liquid, mp -73.96°C, bp 56-57°C. Miscible with water.

Handling - Store at room temperature (25°C).

Safety - Carcinogen/Poison.

Availability - No standard sample supplied.

Current uses - Polyethyleneimine. Used in polyester resins.

Applications - Used as a primer subcoat to anchor epoxy surface coatings to base film/sheet. As an impregnant in the food contact surface of regenerated cellulose film (cellophane) to increase heat sealability. Coating for polyolefin films. As a flocculant used in waste water and potable water treatment.

Analytical methods - Derivatization by reaction with acetyl chloride and determination by capillary GC with nitrogen-selective or electron capture detection.

References - Method under development (Fraunhofer Inst. Munich, D).
J. High Res. Chromatogr., (1989) 12, 604-607.

Ethylene oxide

CAS No.	– 00075-21-8
PM Ref. No.	– 17020
Restrictions	– QM= 1mg/kg in FP.
Formula	– $C_2 H_4 O$
Molecular weight	– 44.05
Alternative names	– Oxirane.

Physical Characteristics – Colourless gas, mp $-111^{O}C$, bp $10.7^{O}C$. Miscible with water, alcohol, and most organic solvents.

Handling – Store at room temperature ($25^{O}C$).

Safety – Toxic/Explosive/Irritant.

Availability – Standard sample supplied, as a solution in toluene at a concentration of 10mg/ml.

Current uses – Polyethylene oxide polymers (polyethers), Co-monomer with terephthalic acid to give polyethylene terephthalate. Co-polymer with starch. Polyester resins.

Applications – Cleaners for ceramic, glass and metal surfaces. Foam stabiliser for malt beverages. Jet cutting for cheese. Bottles. Meat hooks. Paper coatings and adhesives. Wine storage vats. Simulated marble sinks.

Methods of Characterisation – Mass Spectroscopy

Purity – 99.8%

Analytical method – Solid sample heated in sealed vial and headspace sample analysed by GC with flame ionisation detection.

References   – Method under development (Packforsk,
Stockholm, S).
Zh. Anal. Khim., 1984, 34, 100–105.

Ethylene oxide

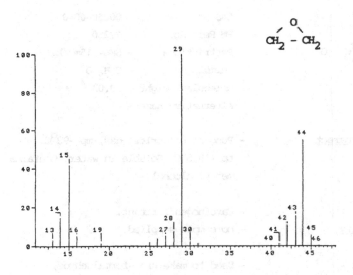

M/Z	Ion Intensity(%)	M/Z	Ion Intensity(%)
13	2.9	30	1.9
14	14.5	31	1.2
15	41.7	40	0.3
16	5.2	41	1.2
19	1.3	42	10.5
25	1.6	43	15.4
26	3.8	44	55.0
27	5.0	45	5.5
28	7.5	46	0.2
29	100.0		

Spectrometer   :Finnigan Mat SSQ 70
Inlet System   :Capillary GC/MS
Source Temperature:150°C
Electron Energy   :70 eV
Scan Range   :25-400

208

Formaldehyde

CAS No.	– 00050–00–0
PM Ref. No.	– 17260
Restrictions	– SML= 15mg/kg
Formula	– CH_2O
Molecular weight	– 30.03
Alternative names–	

H
\
   C = O
/
H

Physical Characteristics – Pungent colourless gas, mp $-92^{\circ}C$ bp $-19.5^{\circ}C$. Soluble in water. Contains methyl alcohol.

Safety – Carcinogen/Irritant.

Availability – No sample supplied.

Current uses – Used to make urea-formaldehyde, melamine-formaldehyde and phenol-formaldehyde resins.

Applications – Used for picnic-ware, kitchen-ware, lining beverage cans for beer and soft drinks.

Analytical methods – In aqueous food simulants by colorimetric method involving reaction with acetylacetone and determination by UV spectrometry. In beverages and simulants by distillation and collection of distillate in a solution of 2,4-dinitrophenylhydrazine. The resulting hydrazone extracted and determined by HPLC with UV detection.

References – Method under development (Packforsk. Stockholm, S).
Food Add. Contam., (1990) 7, 21-27.
Intern. J. Environ. Anal. Chem., (1983) 15, 47-52.

Fumaric acid

CAS No.	– 00110–17–8
PM Ref. No.	– 17290
Restrictions	– none
Formula	– $C_4 H_4 O_4$
Molecular weight	– 116.07
Alternative names	– trans-Butenedioic acid, Boletic acid.

HOOC-HC=CH-COOH

Physical Characteristics	– White powder, mp 299-300°C, sublimes. Soluble in alcohol.
Handling	– Store at room temperature (25°C).
Safety	– Irritant.
Availability	– Standard sample supplied.
Current uses	– Used to make cross-linked polyester resins. Styrene and butadiene polymers (ABS).
Applications	– Storage tanks. Adhesive for co-polymer latexes for bonding polyester fibres and rubbers and coating paper.
Methods of Characterisation	– IR
Purity	– 99%

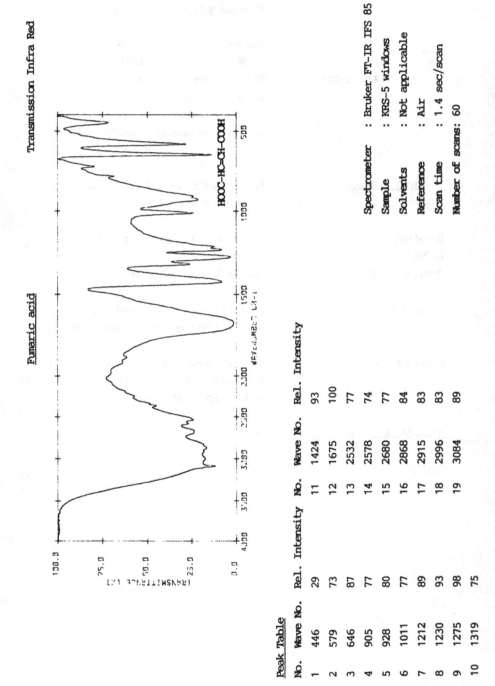

Fumaric acid   Transmission Infra Red

HOOC-HC=CH-COOH

Spectrometer : Bruker FT-IR IFS 85
Sample   : KRS-5 windows
Solvents   : Not applicable
Reference   : Air
Scan time   : 1.4 sec/scan
Number of scans: 60

Peak Table

No.	Wave No.	Rel. Intensity	No.	Wave No.	Rel. Intensity
1	446	29	11	1424	93
2	579	73	12	1675	100
3	646	87	13	2532	77
4	905	77	14	2578	74
5	928	80	15	2680	77
6	1011	77	16	2868	84
7	1212	89	17	2915	83
8	1230	93	18	2996	83
9	1275	98	19	3084	89
10	1319	75			

Glucose

CH$_2$OH

$$\text{(structure)}$$

CAS No.	– 00050–99–7
PM Ref. No.	– 17530
Restrictions	– none
Formula	– C$_6$ H$_{12}$ O$_6$
Molecular weight	– 180.16
Alternative names	– Dextrose, Glucolin.

Physical Characteristics — White powder, mp 146°C. Soluble in water.

Handling — Store at room temperature (25°C).

Safety — Irritant.

Availability — Standard sample supplied.

Current uses — Feedstock for the production of degradable polymers, obtained by means of bacterial fermentation (Polyhydroxybutyrate). Activator for graft polymerisation of acrylonitrile and styrene with rubber. Polymers with phenol and urea.

Applications — Films and containers. Adhesives.

Methods of Characterisation — IR

Purity — 99%

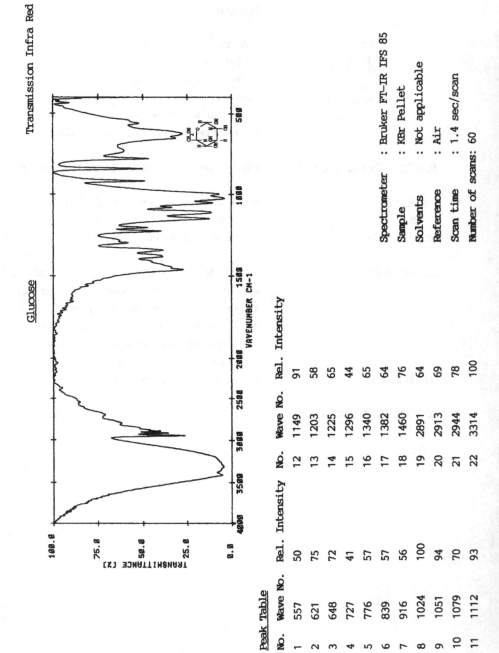

Transmission Infra Red

Glucose

Spectrometer	: Bruker FT-IR IFS 85
Sample	: KBr Pellet
Solvents	: Not applicable
Reference	: Air
Scan time	: 1.4 sec/scan
Number of scans: 60	

Peak Table

No.	Wave No.	Rel. Intensity	No.	Wave No.	Rel. Intensity
1	557	50	12	1149	91
2	621	75	13	1203	58
3	648	72	14	1225	65
4	727	41	15	1296	44
5	776	57	16	1340	65
6	839	57	17	1382	64
7	916	56	18	1460	76
8	1024	100	19	2891	64
9	1051	94	20	2913	69
10	1079	70	21	2944	78
11	1112	93	22	3314	100

Glutaric acid

$$HOOC-(CH_2)_3COOH$$

CAS No.	— 00110-94-1
PM Ref. No.	— 18010
Restrictions	— none
Formula	— $C_5 H_8 O_4$
Molecular weight	— 132.12
Alternative names	— Pentanedioic acid.

Physical Characteristics — White crystaline solid, mp 97°C, bp 302-304°C. Soluble in water, alcohol, ether and chloroform.

Handling — Store at room temperature (25°C).

Safety — Irritant.

Availability — Standard sample supplied.

Current uses — Used to manufacture alkyd resins. Blowing agent for acrylonitrile-styrene co-polymers. Polyesters and polyamides.

Applications — Carbonated drink bottles. Kitchen appliances. Food processing machinery. Coatings.

Methods of Characterisation — IR

Mass Spectroscopy

Purity — 99%

Glutaric acid

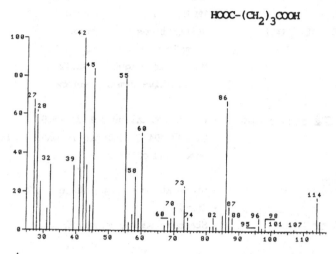

$HOOC-(CH_2)_3COOH$

M/Z	Ion Intensity(%)	M/Z	Ion Intensity(%)
27	62.6	70	6.8
28	59.6	73	21.2
32	33.7	74	4.3
39	33.7	82	2.5
42	100.0	86	63.9
45	79.1	87	7.3
55	74.9	88	2.4
58	27.8	96	2.7
60	48.0	98	1.0
68	5.4	114	15.0

Spectrometer   :Finnigan Mat SSQ 70
Inlet System   :Capillary GC/MS
Source Temperature:150°C
Electron Energy   :70 eV
Scan Range   :25-400

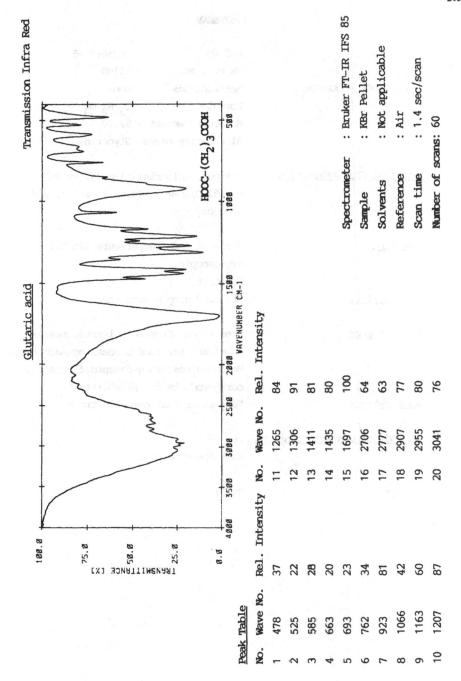

Glutaric acid   Transmission Infra Red

HOOC-(CH₂)₃COOH

WAVENUMBER CM-1

TRANSMITTANCE [%]

Spectrometer	: Bruker FT-IR IFS 85
Sample	: KBr Pellet
Solvents	: Not applicable
Reference	: Air
Scan time	: 1.4 sec/scan
Number of scans: 60	

Peak Table

No.	Wave No.	Rel. Intensity	No.	Wave No.	Rel. Intensity
1	478	37	11	1265	84
2	525	22	12	1306	91
3	585	28	13	1411	81
4	663	20	14	1435	80
5	693	23	15	1697	100
6	762	34	16	2706	64
7	923	81	17	2777	63
8	1066	42	18	2907	77
9	1163	60	19	2955	80
10	1207	87	20	3041	76

Glycerol

OH–CH$_2$–CHOH–CH$_2$OH

CAS No.	– 00056–81–5
PM Ref. No.	– 18100
Restrictions	– none
Formula	– C$_3$ H$_8$ O$_3$
Molecular weight	– 92.10
Alternative names	– Glycerin.

Physical Characteristics
 – Viscous, colourless liquid, mp 20°C, bp 182°C/20mm. Soluble in water, and alcohol.

Handling
 – Store at room temperature (25°C). Hygroscopic.

Safety
 – Irritant.

Availability
 – Standard sample supplied.

Current uses
 – Used to manufacture polyester resins. Lubricant for food processing machinery. Meat casings. In greaseproof paper. In cork products for pliability.

Applications
 – For use as dual ovenable trays.

Methods of Characterisation – IR
 Mass Spectroscopy

Purity
 – 99.5%

Glycerol

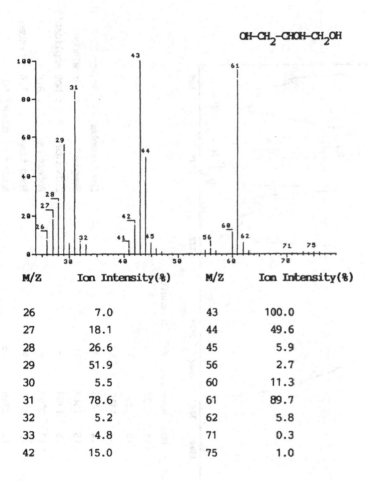

OH–CH$_2$–CHOH–CH$_2$OH

M/Z	Ion Intensity(%)	M/Z	Ion Intensity(%)
26	7.0	43	100.0
27	18.1	44	49.6
28	26.6	45	5.9
29	51.9	56	2.7
30	5.5	60	11.3
31	78.6	61	89.7
32	5.2	62	5.8
33	4.8	71	0.3
42	15.0	75	1.0

Spectrometer   :Finnigan Mat SSQ 70
Inlet System   :Capillary GC/MS
Source Temperature:150°C
Electron Energy   :70 eV
Scan Range   :25–400

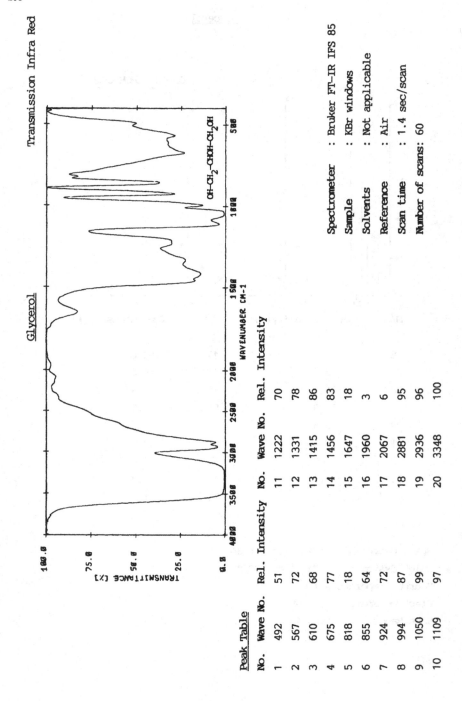

Glycerol   Transmission Infra Red

$OH-CH_2-CHOH-CH_2-OH$

Spectrometer	: Bruker FT-IR IFS 85
Sample	: KBr windows
Solvents	: Not applicable
Reference	: Air
Scan time	: 1.4 sec/scan
Number of scans: 60	

Peak Table

No.	Wave No.	Rel. Intensity	No.	Wave No.	Rel. Intensity
1	492	51	11	1222	70
2	567	72	12	1331	78
3	610	68	13	1415	86
4	675	77	14	1456	83
5	818	18	15	1647	18
6	855	64	16	1960	3
7	924	72	17	2067	6
8	994	87	18	2881	95
9	1050	99	19	2936	96
10	1109	97	20	3348	100

Hexachloroendomethylenetetrahydrophthalic acid

CAS No.	- 00115-28-6
PM Ref. No.	- 18250
Restrictions	- QM= 1mg/kg
Formula	- $C_9 H_4 Cl_6 O_4$
Molecular weight	- 389.00
Alternative names	- Chlorendic acid.

Physical Characteristics — Crystalline solid. Decomposes to form anhydride. Soluble in toluene.

Safety —

Availability — No sample supplied.

Current uses — In the manufacture of unsaturated polyester resins.

Applications — Confers flame retardant qualities.

1-Hexadecanol

$CH_3-(CH_2)_{14}-CH_2OH$

CAS No.	– 36653–82–4
PM Ref. No.	– 18310
Restrictions	– none
Formula	– $C_{16} H_{34} O$
Molecular weight	– 242.45
Alternative names	– Cetyl alcohol.

Physical Characteristics — White crystaline solid, mp 54–56°C, bp 344°C. Soluble in alcohol and chloroform.

Handling — Store at room temperature (25°C).

Safety — Irritant.

Availability — Standard sample supplied.

Current uses — Resins. Defoaming agent. Cellophane.

Applications — Coatings. Sheeting. Mould release agent, lubricant, stabiliser and plasticiser.

Methods of Characterisation — IR

Mass Spectroscopy

Purity — 99%

1-Hexadecanol

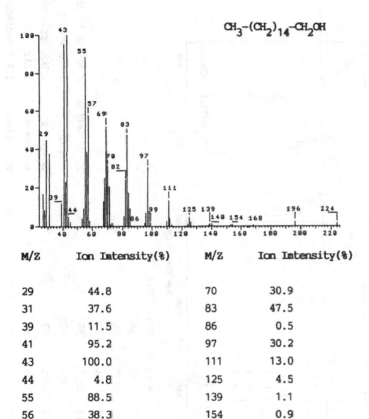

$$CH_3-(CH_2)_{14}-CH_2OH$$

M/Z	Ion Intensity(%)	M/Z	Ion Intensity(%)
29	44.8	70	30.9
31	37.6	83	47.5
39	11.5	86	0.5
41	95.2	97	30.2
43	100.0	111	13.0
44	4.8	125	4.5
55	88.5	139	1.1
56	38.3	154	0.9
57	57.9	196	2.3
69	51.7	224	1.5

Spectrometer   :Finnigan Mat SSQ 70
Inlet System   :Capillary GC/MS
Source Temperature:150°C
Electron Energy   :70 eV
Scan Range   :25–400

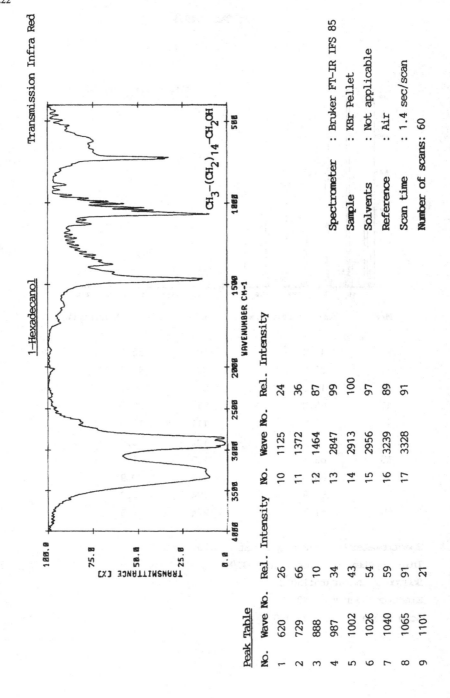

Transmission Infra Red

1-Hexadecanol

$CH_3-(CH_2)_{14}-CH_2OH$

Spectrometer	:	Bruker FT-IR IFS 85
Sample	:	KBr Pellet
Solvents	:	Not applicable
Reference	:	Air
Scan time	:	1.4 sec/scan
Number of scans:		60

Peak Table

No.	Wave No.	Rel. Intensity	No.	Wave No.	Rel. Intensity
1	620	26	10	1125	24
2	729	66	11	1372	36
3	888	10	12	1464	87
4	987	34	13	2847	99
5	1002	43	14	2913	100
6	1026	54	15	2956	97
7	1040	59	16	3239	89
8	1065	91	17	3328	91
9	1101	21			

Hexamethylenediamine

$H_2N-(CH_2)_6-NH_2$

CAS No.	– 00124-09-4
PM Ref. No.	– 18460
Restrictions	– SML=2.4 mg/kg
Formula	– $C_6 H_{16} N_2$
Molecular weight	– 116.21
Alternative names	– 1,6-Hexanediamine, 1,6-Diaminohexane.

Physical Characteristics	– Colourless liquid, mp 42°C, bp 205°C. Soluble in water, alcohol and benzene.
Handling	– Store at room temperature (25°C). Well ventillated.
Safety	– Corrosive/Combustible/Irritant.
Availability	– Standard sample supplied as 80% aqueous solution.
Current uses	– Nylon 6'6. Urethane coatings, co-polymer with sebacic acid (nylon 6/10). Curing agent for epoxy resins. Co-polymer with isophthalic acid (Novamid X21, Gelon A-100).
Applications	– Vacuum packs, gas flushed packs and boil-in-bags. Packaging for meat, fish, coffee and snack foods. Monolayer bottles and films. Refillable soft drink and water bottles. Packaging for nuts.
Methods of Characterisation	– IR Mass Spectroscopy

<u>Purity</u>   - 99%

<u>Analytical methods</u>   - Derivatization with ethylchloroformate in
                                 a two-phase system to form
                                 hexamethylenediamine-diurethane determined
                                 by capillary GC with nitrogen-selective
                                 detection. Alternatively derivatization
                                 with heptafluorobutyric anhydride and
                                 determination of the amides by GC with
                                 nitrogen selective or GC/MS (selected ion
                                 monitoring).

<u>References</u>   - Method under development (Fraunhofer
                                 Inst., Munich, D).
                                 <u>J. Chromatogr.</u>, (1989) <u>479</u>, 125-133.
                                 <u>J. Chromatogr.</u>, (1990) <u>516</u>, 405-413.

Hexamethylenediamine

$H_2N-(CH_2)_6-NH_2$

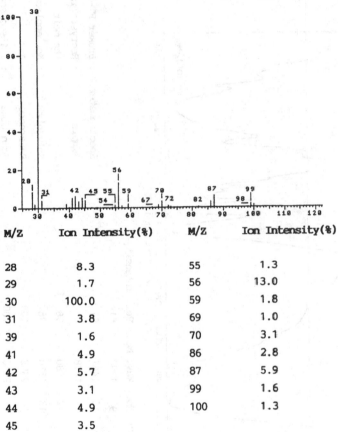

M/Z	Ion Intensity(%)	M/Z	Ion Intensity(%)
28	8.3	55	1.3
29	1.7	56	13.0
30	100.0	59	1.8
31	3.8	69	1.0
39	1.6	70	3.1
41	4.9	86	2.8
42	5.7	87	5.9
43	3.1	99	1.6
44	4.9	100	1.3
45	3.5		

Spectrometer   :Finnigan Mat SSQ 70
Inlet System   :Capillary GC/MS
Source Temperature:150°C
Electron Energy   :70 eV
Scan Range   :25-400

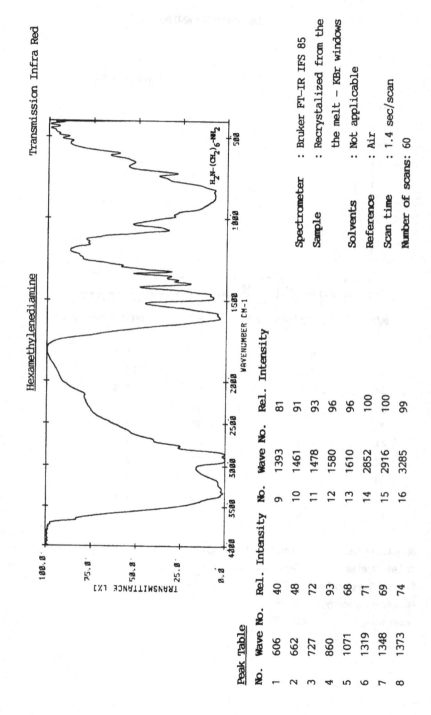

Hexamethylenediamine   Transmission Infra Red

$H_2N-(CH_2)_6-NH_2$

WAVENUMBER CM-1

TRANSMITTANCE [%]

Spectrometer	: Bruker FT-IR IFS 85
Sample	: Recrystalized from the
	the melt – KBr windows
Solvents	: Not applicable
Reference	: Air
Scan time	: 1.4 sec/scan
Number of scans: 60	

Peak Table

No.	Wave No.	Rel. Intensity	No.	Wave No.	Rel. Intensity
1	606	40	9	1393	81
2	662	48	10	1461	91
3	727	72	11	1478	93
4	860	93	12	1580	96
5	1071	68	13	1610	96
6	1319	71	14	2852	100
7	1348	69	15	2916	100
8	1373	74	16	3285	99

Hexamethylene diisocyanate

OCN-(CH$_2$)$_6$-NCO

CAS No.	– 00822-06-0
PM Ref. No.	– 18640
Restrictions	– QM(T)= 1mg/kg in FP (expressed as NCO).
Formula	– C$_8$ H$_{12}$ N$_2$ O$_2$
Molecular weight	– 168.20
Alternative names	– 1,6-Diisocyanatohexane.

Physical Characteristics
- Colourless liquid, mp -67°C, bp 255°C. Soluble in alcohol.

Handling
- Refrigerate (4°C). Protect from moisture.

Safety
- Highly toxic/Irritant.

Availability
- Standard sample supplied.

Current uses
- Used in the manufacture of polyurethanes. and polyamides.

Applications
- Polyurethane tubing for food manufacturing applications. Used to make adhesives in seals for thin films, in polyester paperboard laminates (e.g susceptors) and in multi-layer high barrier plastics (e.g shelf stables) and 'boil-in-the-bag' laminates. Coatings.

Methods of Characterisation - IR
Mass Spectroscopy

Purity
- 98%

Analytical methods   - Isocyanates in materials and articles are analysed by solvent extraction with ethanol in toluene with concurrent urethane derivative formation, clean-up by liquid/liquid partition and solid phase cartridge chromatography and determination by capillary GC with nitrogen selective detection. Phenyl isocyanate and 1,4-butanediisocyanate are used as internal standards.

References   - Draft CEN Method (MAFF, FScL. Norwich, UK).

Hexamethylene diisocyanate

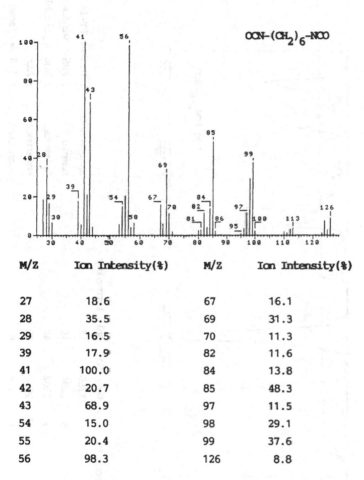

OCN-(CH$_2$)$_6$-NCO

M/Z	Ion Intensity(%)	M/Z	Ion Intensity(%)
27	18.6	67	16.1
28	35.5	69	31.3
29	16.5	70	11.3
39	17.9	82	11.6
41	100.0	84	13.8
42	20.7	85	48.3
43	68.9	97	11.5
54	15.0	98	29.1
55	20.4	99	37.6
56	98.3	126	8.8

Spectrometer   :Finnigan Mat SSQ 70
Inlet System   :Capillary GC/MS
Source Temperature :150°C
Electron Energy   :70 eV
Scan Range   :25-400

Hexamethylene diisocyanate

Transmission Infra Red

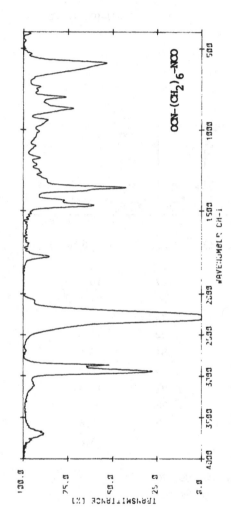

$OCN-(CH_2)_6-NCO$

Spectrometer	: Bruker FT-IR IFS 85
Sample	: KRS-5 windows
Solvents	: Not applicable
Reference	: Air
Scan time	: 1.4 sec/scan
Number of scans: 60	

Peak Table

No.	Wave No.	Rel. Intensity	No.	Wave No.	Rel. Intensity
1	585	46	7	1769	15
2	794	24	8	2284	100
3	864	28	9	2862	48
4	1356	57	10	2940	72
5	1434	23	11	3114	6
6	1464	39	12	3684	12

Hexamethylenetetramine

CAS No.	– 00100–97–0
PM Ref. No.	– 18670
Restrictions	– SML(T)= 15mg/kg (as formaldehyde).
Formula	– $C_6 H_{12} N_4$
Molecular weight	– 140.19
Alternative names	– Methenamine, Hexamine Cystamin, Formin.

Physical Characteristics – White crystals, sublimes 230–270°C. Soluble in water, alcohol, acetone and chloroform.

Handling – Store at room temperature (25°C). Moisture sensitive.

Safety – Flammable/Irritant.

Availability – Standard sample supplied.

Current uses – Antioxidant for polyolefins and polystyrene. Adhesive for metal/rubber laminates. Phenolic thermosetting resins. Cross-linking and curing agent for phenol-formaldehyde resins. Polymerisation control agent for melamine/formaldehyde resins.

Applications – Adhesives, sealing compounds and coatings. Closures.

Methods of Characterisation – IR
Mass Spectroscopy

Purity – 99%

Analytical methods   - Hexamethylenetetramine is treated with acid or heated to release formaldehyde which is reacted with 2,4-dinitrophenylhydrazine. The 2,4-dinitrophenylhydrazone is determined by HPLC with UV detection.

References   - _Analyst_ (1988) <u>113,</u> 511-513.

Hexamethylenetetramine

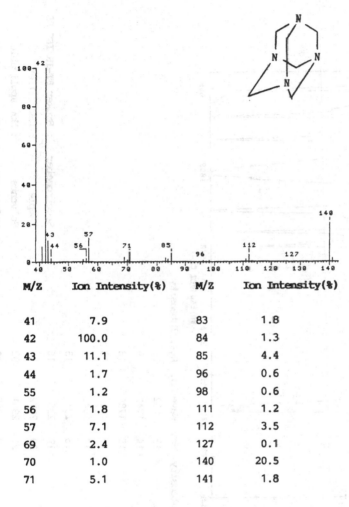

M/Z	Ion Intensity(%)	M/Z	Ion Intensity(%)
41	7.9	83	1.8
42	100.0	84	1.3
43	11.1	85	4.4
44	1.7	96	0.6
55	1.2	98	0.6
56	1.8	111	1.2
57	7.1	112	3.5
69	2.4	127	0.1
70	1.0	140	20.5
71	5.1	141	1.8

Spectrometer   :Finnigan Mat SSQ 70
Inlet System   :Capillary GC/MS
Source Temperature:150°C
Electron Energy   :70 eV
Scan Range   :25–400

234

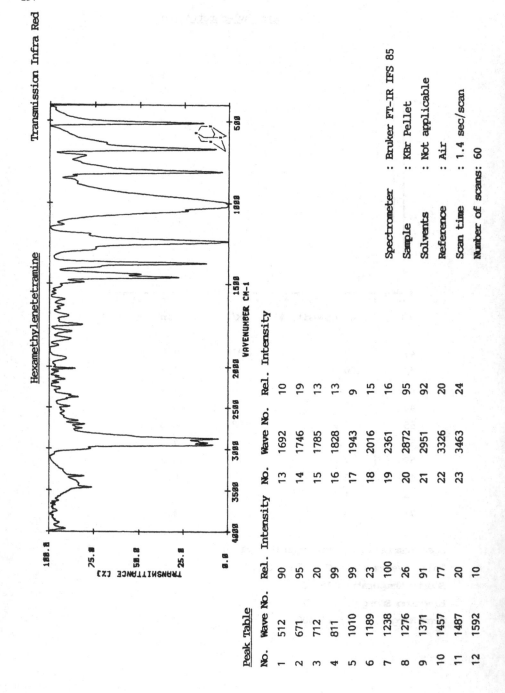

Hexamethylenetetramine

Transmission Infra Red

Peak Table

No.	Wave No.	Rel. Intensity	No.	Wave No.	Rel. Intensity
1	512	90	13	1692	10
2	671	95	14	1746	19
3	712	20	15	1785	13
4	811	99	16	1828	13
5	1010	99	17	1943	9
6	1189	23	18	2016	15
7	1238	100	19	2361	16
8	1276	26	20	2872	95
9	1371	91	21	2951	92
10	1457	77	22	3326	20
11	1487	20	23	3463	24
12	1592	10			

Spectrometer : Bruker FT-IR IFS 85
Sample   : KBr Pellet
Solvents   : Not applicable
Reference   : Air
Scan time   : 1.4 sec/scan
Number of scans: 60

p-Hydroxybenzoic acid

COOH

OH

CAS No.	– 00099-96-7
PM Ref. No.	– 18880
Restrictions	– none
Formula	– $C_7 H_6 O_3$
Molecular weight	– 138.12
Alternative names–	

Physical Characteristics — Pale yellow powder, mp 213-214°C.
Soluble in methanol, ether and acetone.

Handling — Store at room temperature (25°C).

Safety — Irritant.

Availability — Standard sample supplied.

Current uses — Converted to epoxy resin by reaction with epichlorohydrin. Cross-linking agent in phenol-formaldehyde resins. Used as starting substance for ester synthesis. Intermediate for dyes.

Applications — Preservative for adhesives. Coatings.

Methods of Characterisation — IR
Mass Spectroscopy

Purity — 98%

236

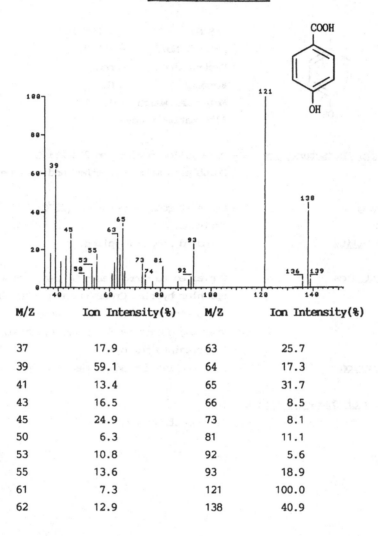

p-Hydroxybenzoic acid

M/Z	Ion Intensity(%)	M/Z	Ion Intensity(%)
37	17.9	63	25.7
39	59.1	64	17.3
41	13.4	65	31.7
43	16.5	66	8.5
45	24.9	73	8.1
50	6.3	81	11.1
53	10.8	92	5.6
55	13.6	93	18.9
61	7.3	121	100.0
62	12.9	138	40.9

Spectrometer   :Finnigan Mat SSQ 70
Inlet System   :Capillary GC/MS
Source Temperature:150°C
Electron Energy   :70 eV
Scan Range   :25–400

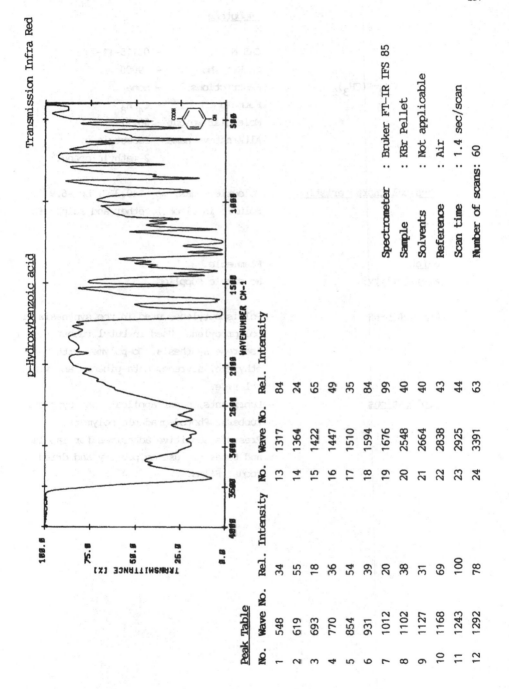

p-Hydroxybenzoic acid

Transmission Infra Red

Peak Table

No.	Wave No.	Rel. Intensity	No.	Wave No.	Rel. Intensity
1	548	34	13	1317	84
2	619	55	14	1364	24
3	693	18	15	1422	65
4	770	36	16	1447	49
5	854	54	17	1510	35
6	931	39	18	1594	84
7	1012	20	19	1676	99
8	1102	38	20	2548	40
9	1127	31	21	2664	40
10	1168	69	22	2838	43
11	1243	100	23	2925	44
12	1292	78	24	3391	63

Spectrometer : Bruker FT-IR IFS 85
Sample : KBr Pellet
Solvents : Not applicable
Reference : Air
Scan time : 1.4 sec/scan
Number of scans: 60

Isobutene

CH$_2$=C(CH$_3$)$_2$

CAS No.   – 00115–11–7
PM Ref. No.   – 19000
Restrictions   – none
Formula   – C$_4$H$_8$
Molecular weight   – 56.11
Alternative names– Isobutylene,
                   2–methylpropene.

Physical Characteristics

– Colourless gas, mp –140.3°C, bp –6.9°C. Soluble in alcohol, ether and sulphuric acid.

Safety

– Flammable.

Availability

– No sample supplied.

Current uses

– Polyisobutylene, used in the synthesis of polypropylene. Used in butyl rubber. Isoprene synthesis. Co–polymer with ethylene, styrene, beta–pinene, and vinyl chloride.

Applications

– Lubricants, paper applications. Synthetic rubbers. Photodegradable polymers. Pressure sensitive adhesives for labels and tapes for use on poultry and dried foods. Films.

Lauric acid

$CH_3-(CH_2)_{10}-COOH$

CAS No.	– 00143-07-7
PM Ref. No.	– 19470
Restrictions	– none
Formula	– $C_{12} H_{24} O_2$
Molecular weight	– 200.32
Alternative names	– Dodecanoic acid.

Physical Characteristics – White powder, mp 44°C, bp 225°C/0.13 bar. Soluble in alcohol and petroleum ether.

Handling – Store at room temperature (25°C).

Safety – Irritant.

Availability – Standard sample supplied.

Current uses – Co-polymer with hexamethylene diamine to give Nylon 6 and Nylon 12.

Applications – Vacuum packs, gas flushed packs, boil-in-bags. Used for coffee, snack foods and processed meats. Laminates with aluminium for high temperature applications. Lubricant for easy opening polyolefin bottle cap liners.

Methods of Characterisation – IR
Mass Spectroscopy

Purity – 99.5%

Lauric acid

$CH_3-(CH_2)_{10}-COOH$

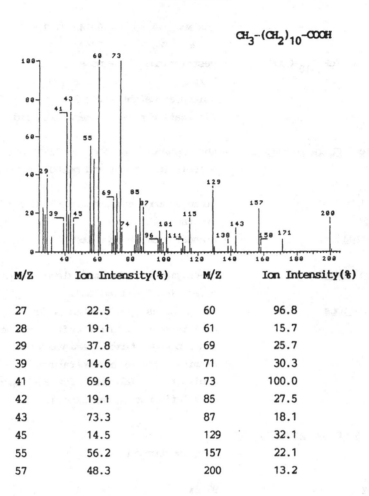

M/Z	Ion Intensity(%)	M/Z	Ion Intensity(%)
27	22.5	60	96.8
28	19.1	61	15.7
29	37.8	69	25.7
39	14.6	71	30.3
41	69.6	73	100.0
42	19.1	85	27.5
43	73.3	87	18.1
45	14.5	129	32.1
55	56.2	157	22.1
57	48.3	200	13.2

Spectrometer   :Finnigan Mat SSQ 70
Inlet System   :Capillary GC/MS
Source Temperature:150°C
Electron Energy   :70 eV
Scan Range   :25–400

Lauric acid, methyl ester

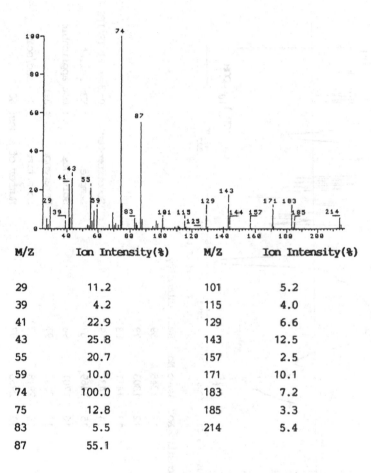

M/Z	Ion Intensity(%)	M/Z	Ion Intensity(%)
29	11.2	101	5.2
39	4.2	115	4.0
41	22.9	129	6.6
43	25.8	143	12.5
55	20.7	157	2.5
59	10.0	171	10.1
74	100.0	183	7.2
75	12.8	185	3.3
83	5.5	214	5.4
87	55.1		

Spectrometer   :Finnigan Mat SSQ 70
Inlet System   :Capillary GC/MS
Source Temperature :150°C
Electron Energy   :70 eV
Scan Range   :25-400

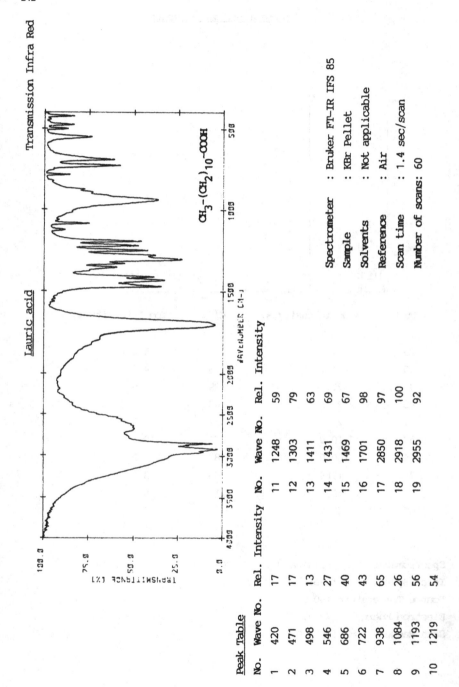

Transmission Infra Red

Lauric acid

$CH_3-(CH_2)_{10}-COOH$

Spectrometer : Bruker FT-IR IFS 85
Sample : KBr Pellet
Solvents : Not applicable
Reference : Air
Scan time : 1.4 sec/scan
Number of scans: 60

Peak Table

No.	Wave No.	Rel. Intensity	No.	Wave No.	Rel. Intensity
1	420	17	11	1248	59
2	471	17	12	1303	79
3	498	13	13	1411	63
4	546	27	14	1431	69
5	686	40	15	1469	67
6	722	43	16	1701	98
7	938	65	17	2850	97
8	1084	26	18	2918	100
9	1193	56	19	2955	92
10	1219	54			

Maleic acid

CAS No.	$-$ 0110–16–7
PM Ref. No.	$-$ 19540
Restrictions	$-$ SML(T)=30 mg/kg
HOOC–CH=CH–COOH	Formula	$-$ $C_4 H_4 O_4$
(cis)	Molecular weight $-$ 116.07
Alternative names– Maleamic acid,
	butenedioic acid,
	cis–1,2–ethylene–
	dicarboxylic acid.

Physical Characteristics	$-$ White crystals, mp 138–139°C. Soluble in water, alcohol and acetone.
Handling	$-$ Store at room temperature (25°C).
Safety	$-$ Corrosive/Irritant.
Availability	$-$ Standard sample supplied.
Current uses	$-$ Cross–linking agent for styrene based polyester resins. Co–polymer for acrylonitrile–butadiene styrene. Co–polymer with ethylene. Co–polymer with styrene for expanded polystyrene & high impact polystyrene. Modifier for vinyl coatings to enhance adhesion to steel or aluminium.
Applications	$-$ Tie layers. Tubs, storage tanks, containers and film. Lacquers.
Methods of Characterisation	$-$ IR
Purity	$-$ 99%
Analytical methods	$-$
References	$-$ Method under development (PIRA, Leatherhead, UK).

Maleic acid

Transmission Infra Red

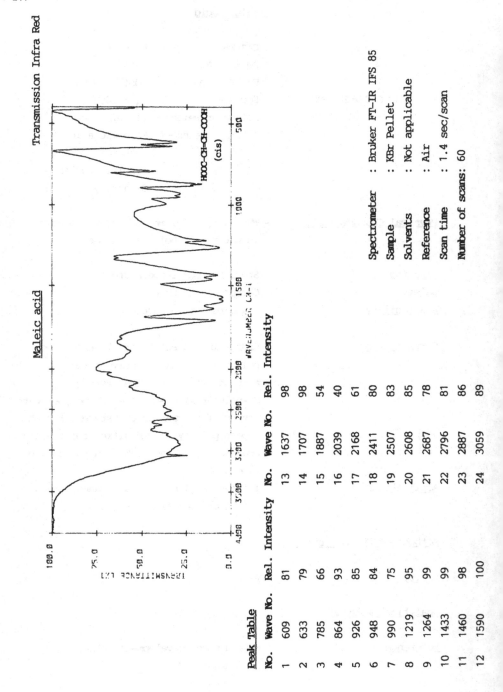

HOOC-CH=CH-COOH
(cis)

Spectrometer	: Bruker FT-IR IFS 85
Sample	: KBr Pellet
Solvents	: Not applicable
Reference	: Air
Scan time	: 1.4 sec/scan
Number of scans: 60	

Peak Table

No.	Wave No.	Rel. Intensity	No.	Wave No.	Rel. Intensity
1	609	81	13	1637	98
2	633	79	14	1707	98
3	785	66	15	1887	54
4	864	93	16	2039	40
5	926	85	17	2168	61
6	948	84	18	2411	80
7	990	75	19	2507	83
8	1219	95	20	2608	85
9	1264	99	21	2687	78
10	1433	99	22	2796	81
11	1460	98	23	2887	86
12	1590	100	24	3059	89

Maleic anhydride

CAS No.   – 0108-31-6
PM Ref. No.   – 19960
Restrictions   – SML(T)= 30 mg/kg
Formula   – $C_4 H_2 O_3$
Molecular weight   – 98.06
Alternative names – 2,5-Furandione.

Physical Characteristics – White solid/powder, mp 52.8°C, bp 202°C.
Soluble in water to give maleic acid.
Soluble in ether, acetone and chloroform.

Handling   – Store at room temperature (25°C). Moisture
sensitive.
Safety   – Corrosive/Toxic.
Availability   – Standard sample supplied.

Current uses   – Polyester resins. Alkyd-type resins.
Co-polymers with ethylene, and propylene.
Co-monomer with styrene to form high
impact polystyrene. Co-monomer in the
manufacture of polyurethane resins. Curing
agent.
Applications   – Tie layers. Tubs, containers, pot lids and
film. Coatings.

Methods of Characterisation – IR

Purity   – 99%

Analytical methods – Maleic anhydride is hydrolysed to the acid
in aqueous simulants. An aliquot of the
aqueous simulant is blown to dryness at
50°C, treated with boron

trifluoride/ethanol and the diethyl
derivative (butenedioic acid diethyl
ester) is determined by capillary GC with
FID. Olive oil simulant is diluted with
hexane and the anhydride extracted as
maleic acid into water which is then
treated as above.

References   - Method under development (PIRA,
                                        Leatherhead, UK).

Transmission Infra Red

Maleic anhydride

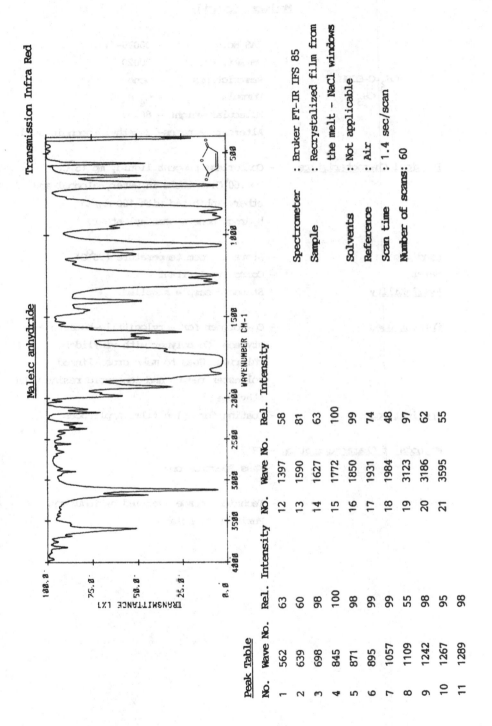

Spectrometer	: Bruker FT-IR IFS 85
Sample	: Recrystalized film from the melt – NaCl windows
Solvents	: Not applicable
Reference	: Air
Scan time	: 1.4 sec/scan
Number of scans: 60	

Peak Table

No.	Wave No.	Rel. Intensity	No.	Wave No.	Rel. Intensity
1	562	63	12	1397	58
2	639	60	13	1590	81
3	698	98	14	1627	63
4	845	100	15	1772	100
5	871	98	16	1850	99
6	895	99	17	1931	74
7	1057	99	18	1984	48
8	1109	55	19	3123	97
9	1242	98	20	3186	62
10	1267	95	21	3595	55
11	1289	98			

Methacrylic acid

$$CH_2=C-COOH$$
$$|$$
$$CH_3$$

CAS No.   – 00079-41-4
PM Ref. No.   – 20020
Restrictions   – none
Formula   – $C_4 H_6 O_2$
Molecular weight   – 86.09
Alternative names – 2-methylpropenoic acid.

Physical Characteristics – Colourless pungent liquid, mp 15°C, bp 160°C. Soluble in water, alcohol and ether. Inhibited with 100 mg/kg hydroquinone monomethyl ether.

Handling   – Store at room temperature (25°C).
Safety   – Corrosive/Irritant.
Availability   – Standard sample supplied.

Current uses   – Co-polymer for acrylonitrile-butadiene-styrene. Co-polymer with vinylidene chloride. Used to make cross-linked polyester resins and ionomeric resins with ethylene.

Applications   – Coating for nylon film. Appliances.

Methods of Characterisation – IR
Mass Spectroscopy

Purity   – Technical grade supplied by industry. Purity not declared.

Methacrylic acid

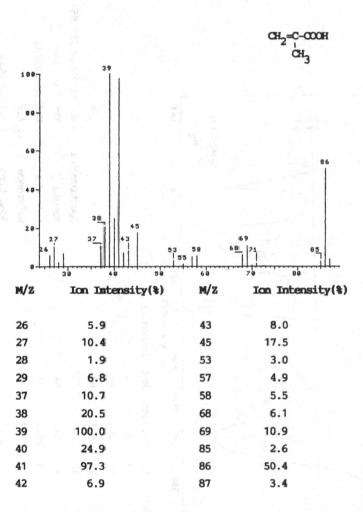

CH$_2$=C-COOH
|
CH$_3$

M/Z	Ion Intensity(%)	M/Z	Ion Intensity(%)
26	5.9	43	8.0
27	10.4	45	17.5
28	1.9	53	3.0
29	6.8	57	4.9
37	10.7	58	5.5
38	20.5	68	6.1
39	100.0	69	10.9
40	24.9	85	2.6
41	97.3	86	50.4
42	6.9	87	3.4

Spectrometer   :Finnigan Mat SSQ 70
Inlet System   :Capillary GC/MS
Source Temperature :150°C
Electron Energy   :70 eV
Scan Range   :25-400

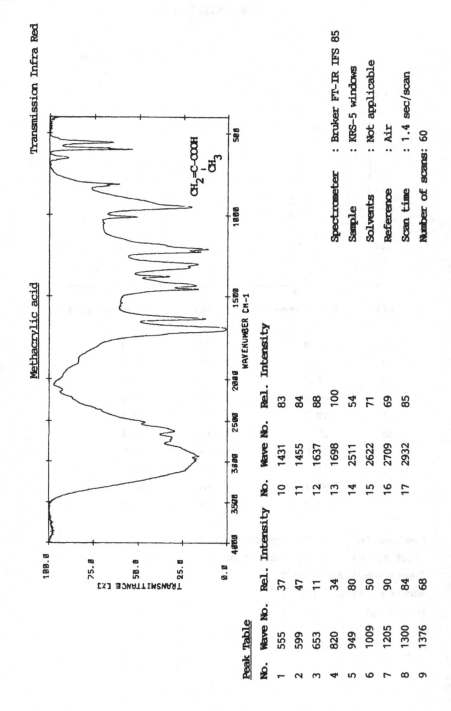

Methacrylic acid   Transmission Infra Red

$CH_2=C-COOH$
$\quad\quad |$
$\quad\quad CH_3$

Spectrometer : Bruker FT-IR IFS 85
Sample   : KRS-5 windows
Solvents   : Not applicable
Reference   : Air
Scan time   : 1.4 sec/scan
Number of scans: 60

Peak Table

No.	Wave No.	Rel. Intensity	No.	Wave No.	Rel. Intensity
1	555	37	10	1431	83
2	599	47	11	1455	84
3	653	11	12	1637	88
4	820	34	13	1698	100
5	949	80	14	2511	54
6	1009	50	15	2622	71
7	1205	90	16	2709	69
8	1300	84	17	2932	85
9	1376	68			

Methacrylic acid, butyl ester

$CH_2=C(CH_3)-COO-(CH_2)_3-CH_3$

CAS No.	– 00097–88–1
PM Ref. No.	– 20110
Restrictions	– none
Formula	– $C_8 H_{14} O_2$
Molecular weight	– 142.2
Alternative names	– n–Butyl methacrylate.

Physical Characteristics — Colourless liquid, bp 160–163°C. Inhibited with 10 mg/kg hydroquinone monomethyl ether.

Handling — Refrigerate (4°C).

Safety — Irritant.

Availability — Standard sample supplied.

Current uses — n–Butyl methacrylate homopolymers. Cross–linking agent for polyester resins. Co–monomer in polyvinylidene chloride, high impact polystyrene, polyvinyl chloride, polyethylene and polystyrene.

Applications — Rigid and semi-rigid acrylic plastics for repeat use. Films and coatings. Adhesives.

Methods of Characterisation — IR
Mass Spectroscopy

Purity — 99%

252

Methacrylic acid, butyl ester

$CH_2=C(CH_3)-COO-(CH_2)_3-CH_3$

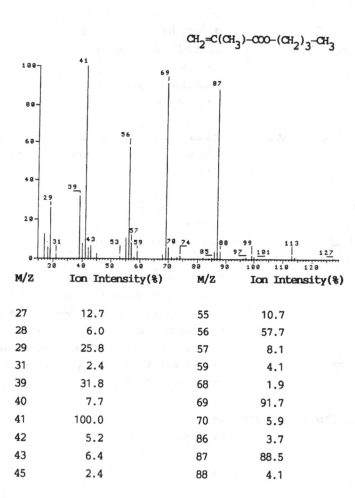

M/Z	Ion Intensity(%)	M/Z	Ion Intensity(%)
27	12.7	55	10.7
28	6.0	56	57.7
29	25.8	57	8.1
31	2.4	59	4.1
39	31.8	68	1.9
40	7.7	69	91.7
41	100.0	70	5.9
42	5.2	86	3.7
43	6.4	87	88.5
45	2.4	88	4.1

Spectrometer   :Finnigan Mat SSQ 70
Inlet System   :Capillary GC/MS
Source Temperature:150°C
Electron Energy   :70 eV
Scan Range   :25–400

Methacrylic acid, butyl ester   Transmission Infra Red

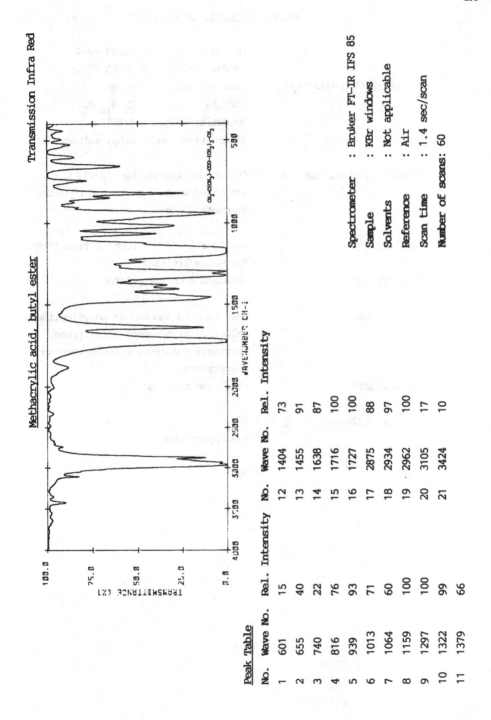

Spectrometer   : Bruker FT-IR IFS 85
Sample   : KBr windows
Solvents   : Not applicable
Reference   : Air
Scan time   : 1.4 sec/scan
Number of scans: 60

Peak Table

No.	Wave No.	Rel. Intensity	No.	Wave No.	Rel. Intensity
1	601	15	12	1404	73
2	655	40	13	1455	91
3	740	22	14	1638	87
4	816	76	15	1716	100
5	939	93	16	1727	100
6	1013	71	17	2875	88
7	1064	60	18	2934	97
8	1159	100	19	2962	100
9	1297	100	20	3105	17
10	1322	99	21	3424	10
11	1379	66			

Methacrylic acid, ethyl ester

$$CH_2=C(CH_3)-COO-CH_2CH_3$$

CAS No.	— 00097–63–2
PM Ref. No.	— 20890
Restrictions	— none
Formula	— $C_6 H_{10} O_2$
Molecular weight	— 114.15
Alternative names	— Ethyl methacrylate.

Physical Characteristics — Colourless liquid, bp 118–119°C. Inhibited with 15 mg/kg hydroquinone monomethyl ether.

Handling — Refrigerate (4°C). Protect from light.

Safety — Flammable/Irritant.

Availability — Standard sample supplied.

Current uses — Used as a co-monomer in polyvinylidene chloride, high impact polystyrene, polyvinyl chloride, polyethylene and polystyrene.

Applications — Films and coatings.

Methods of Characterisation — IR

Mass Spectroscopy

Purity — 99%

Methacrylic acid, ethyl ester

$$CH_2=C(CH_3)-COO-CH_2CH_3$$

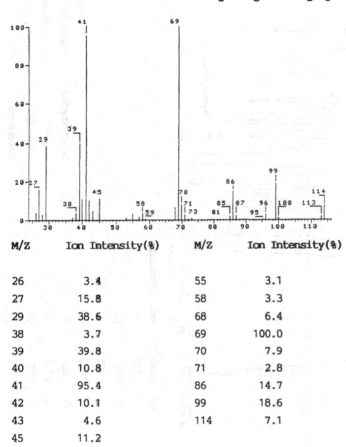

M/Z	Ion Intensity(%)	M/Z	Ion Intensity(%)
26	3.4	55	3.1
27	15.8	58	3.3
29	38.6	68	6.4
38	3.7	69	100.0
39	39.8	70	7.9
40	10.8	71	2.8
41	95.4	86	14.7
42	10.1	99	18.6
43	4.6	114	7.1
45	11.2		

Spectrometer   :Finnigan Mat SSQ 70
Inlet System   :Capillary GC/MS
Source Temperature:150°C
Electron Energy   :70 eV
Scan Range   :25–400

Methacrylic acid, ethyl ester   Transmission Infra Red

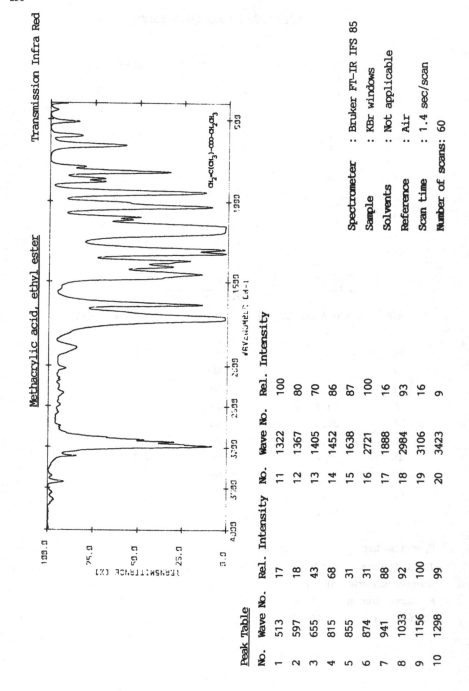

CH$_2$=C(CH$_3$)-COO-CH$_2$CH$_3$

Spectrometer	: Bruker FT-IR IFS 85
Sample	: KBr windows
Solvents	: Not applicable
Reference	: Air
Scan time	: 1.4 sec/scan
Number of scans: 60	

Peak Table

No.	Wave No.	Rel. Intensity	No.	Wave No.	Rel. Intensity
1	513	17	11	1322	100
2	597	18	12	1367	80
3	655	43	13	1405	70
4	815	68	14	1452	86
5	855	31	15	1638	87
6	874	31	16	2721	100
7	941	88	17	1888	16
8	1033	92	18	2984	93
9	1156	100	19	3106	16
10	1298	99	20	3423	9

Methacrylic acid, ethylene glycol monoester

$$OH-CH_2CH_2-O-\overset{\overset{\displaystyle O}{\|}}{C}-\underset{\underset{\displaystyle CH_3}{|}}{C}=CH_2$$

CAS No.	– 00868-77-9
PM Ref. No.	– 21190
Restrictions	– none
Formula	– $C_6 H_{10} O_3$
Molecular weight	– 130.14
Alternative names	– 2-Hydroxyethyl methacrylate.

Physical Characteristics
 – Colourless liquid, mp -60°C, bp 82°C/3 mbar. Soluble in water and most organic solvents. Inhibited with 200 mg/kg p-methoxyphenol.

Handling
 – Refrigerate (4°C), protect from light.

Safety
 – Toxic/Irritant.

Availability
 – Standard sample supplied.

Current uses
 – Used as a co-monomer for vinylidene chloride, high impact polystyrene, polyvinyl acetate, and as a co-polymer in polyethylene. As a co-polymer with butyl acrylate. In graft polymerisations.

Applications
 – Coatings and adhesives.

Methods of Characterisation
 – IR
 Mass Spectroscopy

Purity
 – Technical grade supplied by industry. Purity not declared.

258

Methacrylic acid, ethylene glycol monoester

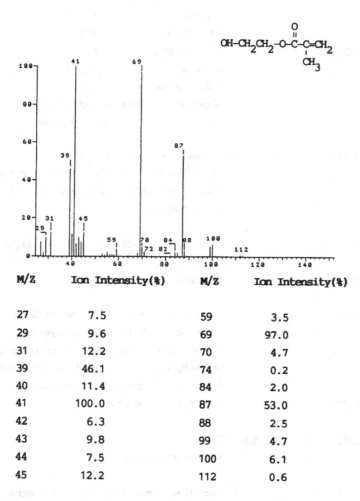

M/Z	Ion Intensity(%)	M/Z	Ion Intensity(%)
27	7.5	59	3.5
29	9.6	69	97.0
31	12.2	70	4.7
39	46.1	74	0.2
40	11.4	84	2.0
41	100.0	87	53.0
42	6.3	88	2.5
43	9.8	99	4.7
44	7.5	100	6.1
45	12.2	112	0.6

Spectrometer   :Finnigan Mat SSQ 70
Inlet System   :Capillary GC/MS
Source Temperature:150°C
Electron Energy   :70 eV
Scan Range   :25-400

259

Methacrylic acid, ethylene glycol monoester   Transmission Infra Red

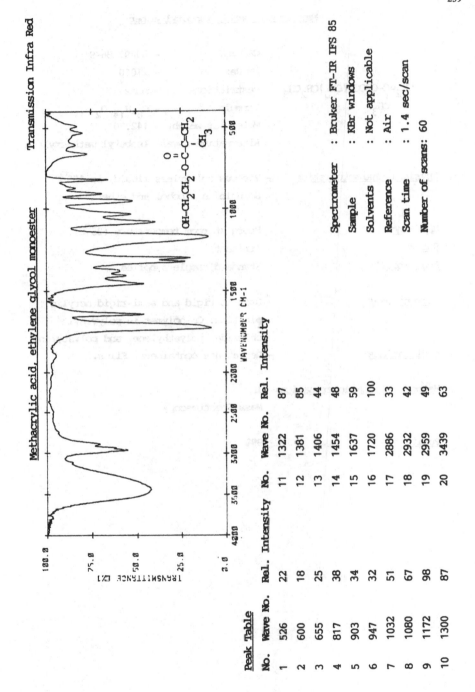

$CH_2-CH_2CH_2-O-C-C=CH_2$
                    $\overset{\displaystyle O}{\|}$ CH_3

Spectrometer   : Bruker FT-IR IFS 85
Sample   : KBr windows
Solvents   : Not applicable
Reference   : Air
Scan time   : 1.4 sec/scan
Number of scans: 60

Peak Table

No.	Wave No.	Rel. Intensity	No.	Wave No.	Rel. Intensity
1	526	22	11	1322	87
2	600	18	12	1381	85
3	655	25	13	1406	44
4	817	38	14	1454	48
5	903	34	15	1637	59
6	947	32	16	1720	100
7	1032	51	17	2886	33
8	1080	67	18	2932	42
9	1172	98	19	2959	49
10	1300	87	20	3439	63

Methacrylic acid, isobutyl ester

$CH_2=\underset{\underset{CH_3}{|}}{C}-COOCH(CH_3)CH_2CH_3$

CAS No.	– 00097–86–9
PM Ref. No.	– 21010
Restrictions	– none
Formula	– $C_8\ H_{14}\ O_2$
Molecular weight	– 142.20
Alternative names	– Isobutyl methacrylate.

Physical Characteristics — Viscous colourless liquid, bp 155°C. Soluble in alcohol and ether.

Handling — Store at room temperature (25°C).

Safety — Irritant.

Availability — Standard sample supplied.

Current uses — Used in rigid and semi-rigid acrylic plastics. Co-polymer in polyvinyl chloride, polyethylene, and polystyrene.

Applications — Repeat-use containers. Films.

Methods of Characterisation — IR

Mass Spectroscopy

Purity — 99%

Methacrylic acid, isobutyl ester

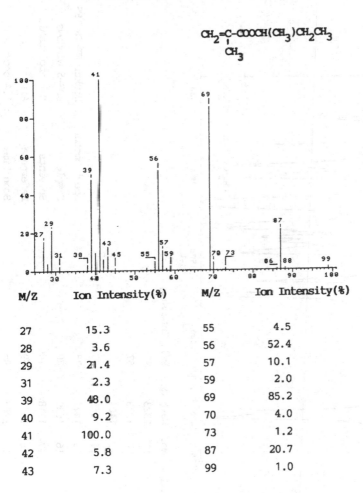

$$CH_2=\underset{\underset{CH_3}{|}}{C}-COOCH(CH_3)CH_2CH_3$$

M/Z	Ion Intensity(%)	M/Z	Ion Intensity(%)
27	15.3	55	4.5
28	3.6	56	52.4
29	21.4	57	10.1
31	2.3	59	2.0
39	48.0	69	85.2
40	9.2	70	4.0
41	100.0	73	1.2
42	5.8	87	20.7
43	7.3	99	1.0

Spectrometer   :Finnigan Mat SSQ 70
Inlet System   :Capillary GC/MS
Source Temperature:150°C
Electron Energy   :70 eV
Scan Range   :25-400

Methacrylic acid, isobutyl ester

Transmission Infra Red

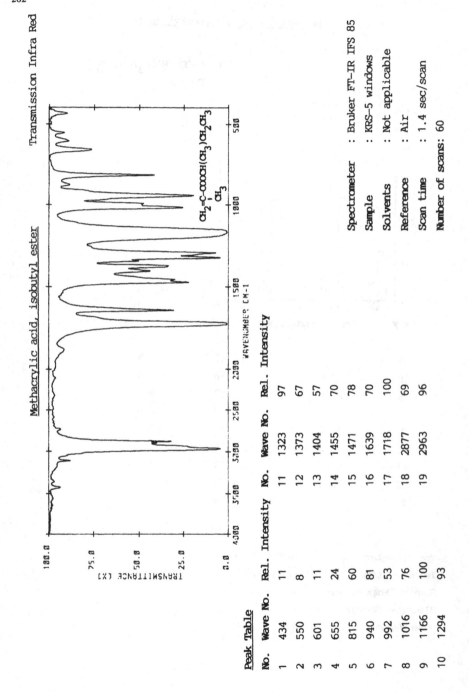

Spectrometer	: Bruker FT-IR IFS 85
Sample	: KRS-5 windows
Solvents	: Not applicable
Reference	: Air
Scan time	: 1.4 sec/scan
Number of scans: 60	

Peak Table

No.	Wave No.	Rel. Intensity	No.	Wave No.	Rel. Intensity
1	434	11	11	1323	97
2	550	8	12	1373	67
3	601	11	13	1404	57
4	655	24	14	1455	70
5	815	60	15	1471	78
6	940	81	16	1639	70
7	992	53	17	1718	100
8	1016	76	18	2877	69
9	1166	100	19	2963	96
10	1294	93			

Methacrylic acid, isopropyl ester

$$CH_2=C-COOCH(CH_3)_2$$
$$\quad\quad\overset{|}{CH_3}$$

CAS No.	– 04655–34–9
PM Ref. No.	– 21100
Restrictions	– none
Formula	– $C_7 H_{12} O_2$
Molecular weight	– 128.17
Alternative names	– Isopropyl methacrylate.

Physical Characteristics – Bp 125°C.

Safety –

Availability – No sample supplied.

Current uses – Used in rigid and semi-rigid acrylic plastics. Used as a co-monomer for polyvinylidene chloride, high impact polystyrene, polyvinyl chloride, polyethylene, and polystyrene.

Applications – Repeat-use containers. Films.

Methacrylic acid, methyl ester

$CH_2=C(CH_3)-COOCH_3$

CAS No.   – 00080–62–6
PM Ref. No.   – 21130
Restrictions   – none
Formula   – $C_5 H_8 O_2$
Molecular weight   – 100.12
Alternative names– Methyl methacrylate.

Physical Characteristics   – Colourless liquid, pungent odour, mp –48°C, bp 100.3°C. Soluble in water, alcohol, ether, and acetone. Polymerises on exposure to light. Inhibited with 15–20 mg/kg of hydroquinone monomethyl ether.

Handling   – Store at room temperature (25°C). Well ventillated.

Safety   – Flammable/Irritant.

Availability   – Standard sample supplied.

Current uses   – Polymethyl methacrylate. Cross–linking agent for unsaturated polyester resins. Terpolymer with acrylonitrile and vinylidene chloride. Co–polymer with ethylene. Used in elastomeric latexes, and acrylate ester co–polymers. Polymer modifier for vinyl chloride plastics.

Applications   – Film & sheeting. Dual ovenable trays. Rigid and semi–rigid repeat–use containers. Lacquers. Coating for nylon film. Shrink wrap.

Methods of Characterisation – IR
Mass Spectroscopy

Purity   – 99.8% wt/wt

Methacrylic acid, methyl ester

$$CH_2=C(CH_3)-COOCH_3$$

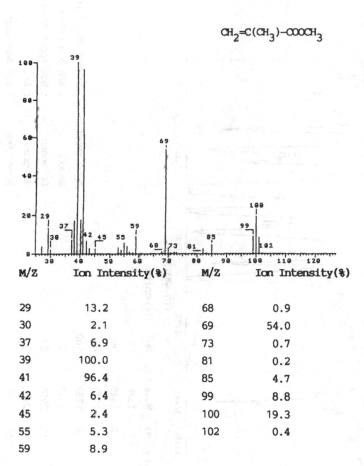

M/Z	Ion Intensity(%)	M/Z	Ion Intensity(%)
29	13.2	68	0.9
30	2.1	69	54.0
37	6.9	73	0.7
39	100.0	81	0.2
41	96.4	85	4.7
42	6.4	99	8.8
45	2.4	100	19.3
55	5.3	102	0.4
59	8.9		

Spectrometer   :Finnigan Mat SSQ 70
Inlet System   :Capillary GC/MS
Source Temperature:150°C
Electron Energy   :70 eV
Scan Range   :25-400

266

Methacrylic acid, methyl ester   Transmission Infra Red

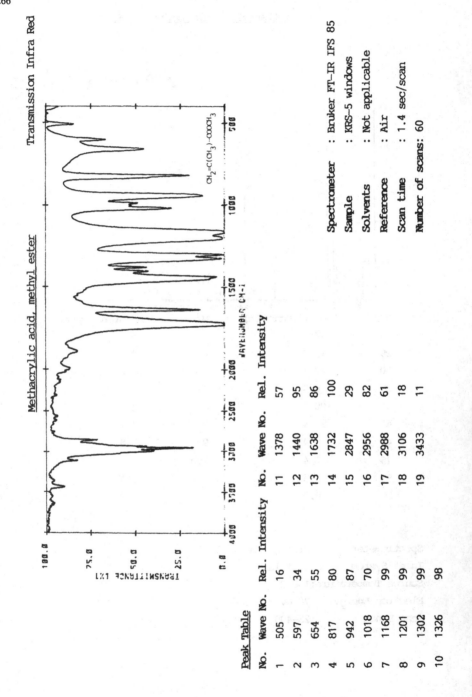

CH₂=C(CH₃)-COOCH₃

Spectrometer	: Bruker FT-IR IFS 85
Sample	: KRS-5 windows
Solvents	: Not applicable
Reference	: Air
Scan time	: 1.4 sec/scan
Number of scans: 60	

Peak Table

No.	Wave No.	Rel. Intensity	No.	Wave No.	Rel. Intensity
1	505	16	11	1378	57
2	597	34	12	1440	95
3	654	55	13	1638	86
4	817	80	14	1732	100
5	942	87	15	2847	29
6	1018	70	16	2956	82
7	1168	99	17	2988	61
8	1201	99	18	3106	18
9	1302	99	19	3433	11
10	1326	98			

Methacrylic acid, propyl ester

$$CH_2=C-COOCH_2CH_2CH_3$$
$$CH_3$$

CAS No.	– 02210-28-8
PM Ref. No.	– 21340
Restrictions	– none
Formula	– $C_7 H_{12} O_2$
Molecular weight	– 128.17
Alternative names	– Propyl methacrylate.

Physical Characteristics   – Bp 141°C

Safety   –

Availability   – No sample supplied.

Current uses   – Used in rigid and semi-rigid acrylic plastics. Used as a co-monomer for polyvinylidene chloride, high impact polystyrene, polyvinyl chloride, polyethylene, and polystyrene.

Applications   – Repeat-use containers. Films.

Methacrylic acid, sec butyl ester

$CH_2=C-COOCH(CH_3)CH_2CH_3$
  $|$
  CH_3

CAS No.	- 02998-18-7
PM Ref. No.	- 20140
Restrictions	- none
Formula	- $C_8 H_{14} O_2$
Molecular weight	- 142.20
Alternative names	-

Physical Characteristics   -

Safety   -

Availability   - No sample supplied.

Current uses   - Used in rigid and semi-rigid acrylic
                              plastics. Used as a co-monomer for
                              polyvinylidene chloride, high impact
                              polystyrene, polyvinyl chloride,
                              polyethylene, and polystyrene.

Applications   - Repeat-use containers. Films.

Methacrylic acid, tert. butyl ester

$$CH_2=C-COOC(CH_3)_3$$
$$CH_3$$

CAS No.	– 00585–07–9
PM Ref. No.	– 20170
Restrictions	– none
Formula	– $C_8 H_{14} O_2$
Molecular weight	– 142.20
Alternative names–	

Physical Characteristics – Bp 52°C.

Safety   –

Availability   – No sample supplied.

Current uses   – Used in rigid and semi-rigid acrylic plastics. Used as a co-monomer for polyvinylidene chloride, high impact polystyrene, polyvinyl chloride, polyethylene, and polystyrene.

Applications   – Repeat-use containers. Films.

Methacrylic anhydride

$$CH_2=C-COOCO-C=CH_2$$
$$\quad CH_3 \qquad CH_3$$

CAS No.   – 00760-93-0
PM Ref. No.   – 21460
Restrictions   – none
Formula   – $C_8 H_{10} O_3$
Molecular weight – 154.17
Alternative names–

Physical Characteristics   – Colourless liquid, bp 87°C/0.02 bar.
                             Soluble in alcohol, and ether. Inhibited
                             by 2000 mg/kg Tocanol A.

Handling   – Store at room temperature (25°C). Protect
                     from moisture.

Safety   – Corrosive.

Availability   – Standard sample supplied (partially
                     hydrolysed to acid).

Current uses   – Used in unsaturated polyesters.

Applications   – Carbonated drinks bottles. Kitchen
                     appliances.

Methods of Characterisation – IR
                              Mass Spectroscopy

Purity   – 70%

Methacrylic anhydride

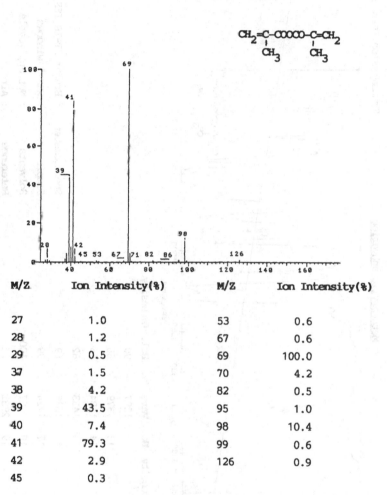

M/Z	Ion Intensity(%)	M/Z	Ion Intensity(%)
27	1.0	53	0.6
28	1.2	67	0.6
29	0.5	69	100.0
37	1.5	70	4.2
38	4.2	82	0.5
39	43.5	95	1.0
40	7.4	98	10.4
41	79.3	99	0.6
42	2.9	126	0.9
45	0.3		

Spectrometer   :Finnigan Mat SSQ 70
Inlet System   :Capillary GC/MS
Source Temperature:150°C
Electron Energy   :70 eV
Scan Range   :25-400

Methacrylic anhydride

Transmission Infra Red

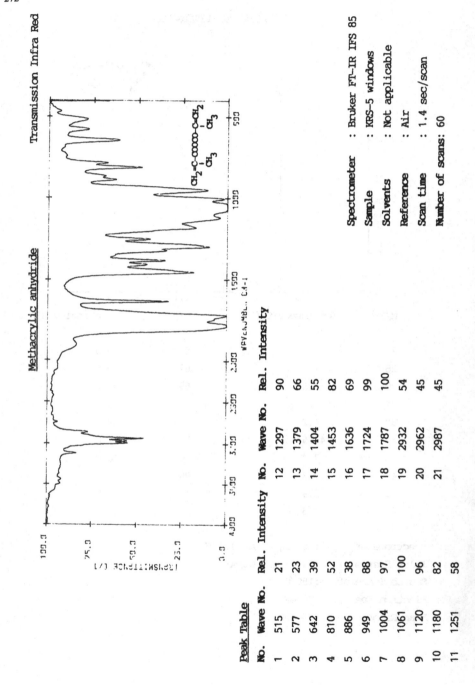

Spectrometer	:	Bruker FT-IR IFS 85
Sample	:	KRS-5 windows
Solvents	:	Not applicable
Reference	:	Air
Scan time	:	1.4 sec/scan
Number of scans: 60		

Peak Table

No.	Wave No.	Rel. Intensity	No.	Wave No.	Rel. Intensity
1	515	21	12	1297	90
2	577	23	13	1379	66
3	642	39	14	1404	55
4	810	52	15	1453	82
5	886	38	16	1636	69
6	949	88	17	1724	99
7	1004	97	18	1787	100
8	1061	100	19	2932	54
9	1120	96	20	2962	45
10	1180	82	21	2987	45
11	1251	58			

Methacrylonitrile

$CH_2=C(CH_3)-CN$

CAS No.	— 00126–98–7
PM Ref. No.	— 21490
Restrictions	— SML= Not detectable
	(DL= 0.02mg/kg)
Formula	— $C_4 H_5 N$
Molecular weight	— 67.09
Alternative names—	

Physical Characteristics — Colourless liquid, mp -35.8^{o}C, bp 90–92°C. Soluble in alcohol and ether. Stabilised with 50 mg/kg hydroquinone monomethyl ether.

Handling — Store at room temperature (25°C).

Safety — Highly Toxic/Irritant.

Availability — Standard sample supplied.

Current uses — Used as an alternative to acrylonitrile as a co-monomer with vinylidene chloride polymers. Co-polymer with styrene (Lopac), and with polybutadiene.

Applications — Coating for regenerated cellulose film, for the packaging biscuits and snack foods. Packaging for dairy and meat products. Bottles for carbonated beverages.

Methods of Characterisation — IR

Mass Spectroscopy

<u>Purity</u>	– 99%
<u>Analytical methods</u>	– Headspace GC with nitrogen selective detection. Foods or simulants equilibrated at 70°C prior to headspace sampling. Proprionitrile used as internal standard. Confirmation by GC/MS selected ion monitoring.
<u>References</u>	– Method under development (Fraunhofer Inst., Munich, D).

Methacrylonitrile

$$CH_2=C(CH_3)-CN$$

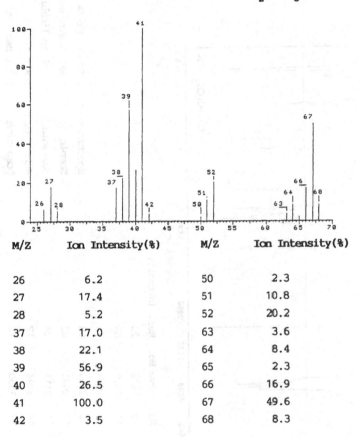

M/Z	Ion Intensity(%)	M/Z	Ion Intensity(%)
26	6.2	50	2.3
27	17.4	51	10.8
28	5.2	52	20.2
37	17.0	63	3.6
38	22.1	64	8.4
39	56.9	65	2.3
40	26.5	66	16.9
41	100.0	67	49.6
42	3.5	68	8.3

Spectrometer   :Finnigan Mat SSQ 70
Inlet System   :Capillary GC/MS
Source Temperature:150°C
Electron Energy   :70 eV
Scan Range   :25–400

Methacrylonitrile

Transmission Infra Red

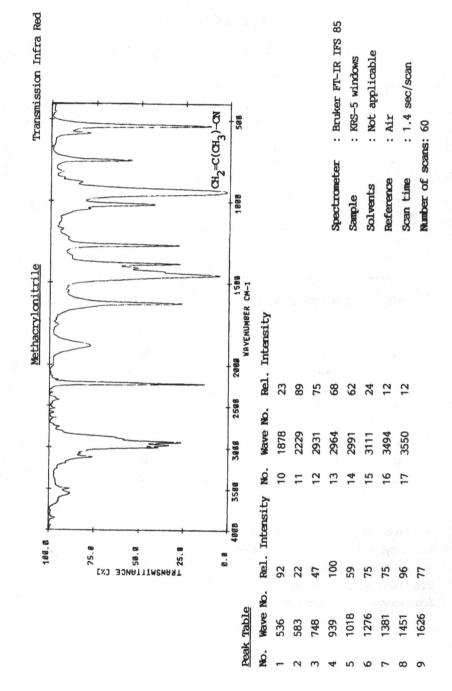

$CH_2=C(CH_3)-CN$

Spectrometer	: Bruker FT-IR IFS 85
Sample	: KRS-5 windows
Solvents	: Not applicable
Reference	: Air
Scan time	: 1.4 sec/scan
Number of scans: 60

Peak Table

No.	Wave No.	Rel. Intensity	No.	Wave No.	Rel. Intensity
1	536	92	10	1878	23
2	583	22	11	2229	89
3	748	47	12	2931	75
4	939	100	13	2964	68
5	1018	59	14	2991	62
6	1276	75	15	3111	24
7	1381	75	16	3494	12
8	1451	96	17	3550	12
9	1626	77			

Methanol

CH₃OH

CAS No.	– 00067–56–1
PM Ref. No.	– 21550
Restrictions	– none
Formula	– C H$_4$ O
Molecular weight	– 32.04
Alternative names	– Methyl alcohol; wood alcohol.

Physical Characteristics – Colourless liquid, mp –97.8°C, bp 64.7°C.

Handling – Store at room temperature (25°C).

Safety – Toxic/Flammable.

Availability – Standard sample supplied.

Current uses – Starting substance for methyl methacrylate; terephthalic acid, dimethyl ester; and formaldehyde. Alkylating agent for amino resins.

Applications – Film and sheeting. Bottles. Picnic-ware. Coatings.

Methods of Characterisation – IR
Mass Spectroscopy

Purity – 99.9%

Methanol

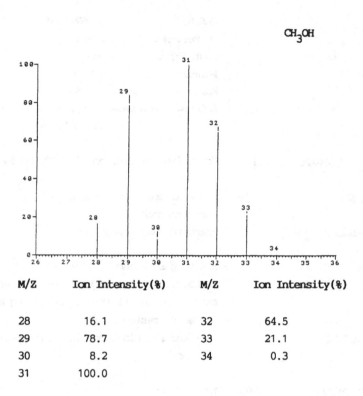

M/Z	Ion Intensity(%)	M/Z	Ion Intensity(%)
28	16.1	32	64.5
29	78.7	33	21.1
30	8.2	34	0.3
31	100.0		

Spectrometer   :Finnigan Mat SSQ 70
Inlet System   :Capillary GC/MS
Source Temperature :150°C
Electron Energy   :70 eV
Scan Range   :25–400

279

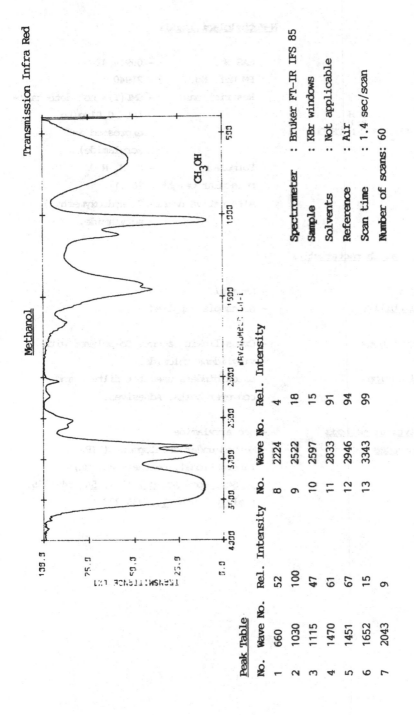

Methanol　　　　　　Transmission Infra Red

CH₃OH

Spectrometer	: Bruker FT-IR IFS 85
Sample	: KBr windows
Solvents	: Not applicable
Reference	: Air
Scan time	: 1.4 sec/scan
Number of scans: 60	

Peak Table

No.	Wave No.	Rel. Intensity	No.	Wave No.	Rel. Intensity
1	660	52	8	2224	4
2	1030	100	9	2522	18
3	1115	47	10	2597	15
4	1470	61	11	2833	91
5	1451	67	12	2946	94
6	1652	15	13	3343	99
7	2043	9			

N-Methylolacrylamide

CH$_2$=CH-C-N-CH$_2$OH (with O double-bonded to C and H on N)

CAS No.	- 00924-42-5
PM Ref. No.	- 21940
Restrictions	- SML(T)= not detectable (DL= 0.01mg/kg expressed as acrylamide).
Formula	- C$_4$ H$_7$ N O$_2$
Molecular weight	- 101.11
Alternative names	- N-(Hydroxymethyl) acrylamide.

Physical Characteristics   -

Safety   - Harmful.
Availability   - No sample supplied.

Current uses   - Cross-linking agent. Co-polymer with vinylidene chloride.
Applications   - Latex binders used for filters and conveyer belts. Adhesives.

Analytical methods - See acrylamide.
References   - Method under development (PIRA International, Leatherhead, UK).
J. Sci. Food Agric., 1991, 54, 549-555.
Analyst, 1988, 13, 335-338.

4-Methyl-1-Pentene

$$CH_3-CH-CH_2-CH=CH_2$$
with CH_3 branch on the second carbon

CAS No.	– 00691-37-2
PM Ref. No.	– 22150
Restrictions	– SML= 0.02mg/kg
Formula	– C_6H_{12}
Molecular weight	– 84.16
Alternative names–	

Physical Characteristics   – Colourless liquid, bp 53-54°C.

Handling   – Refrigerate (4°C).

Safety   – Flammable/Irritant.

Availability   – Standard sample supplied.

Current uses   – Co-monomer with linear low density polyethylene. Outer layer of polyolefin laminates. Polymethylpentene (TPX, PMP).

Applications   – Film for pre-packed fresh & frozen foods. Part of laminate with paper or Al made into bags and blow-moulded for storage containers – good oil resistance. Heat sealable films. Hot filling. Microwaveable packaging and articles.

Methods of Characterisation – IR
Mass Spectroscopy

Purity   – Technical grade supplied by industry. Purity not declared.

Analytical methods   – Direct headspace analysis above simulants using capillary GC and FID (or GC/MS selected ion monitoring).

References   –

282

4-Methyl-1-Pentene

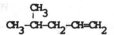

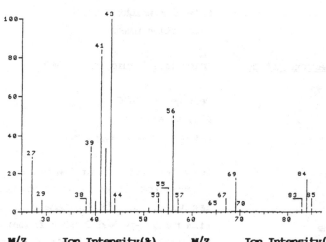

M/Z	Ion Intensity(%)	M/Z	Ion Intensity(%)
27	26.9	55	10.1
29	5.7	56	48.4
38	2.7	57	2.6
39	30.4	65	1.0
40	5.1	67	2.6
41	80.4	69	15.2
42	33.4	70	0.9
43	100.0	83	1.1
44	3.3	84	16.7
53	3.4	85	1.5

Spectrometer   :Finnigan Mat SSQ 70
Inlet System   :Capillary GC/MS
Source Temperature:150°C
Electron Energy   :70 eV
Scan Range   :25-400

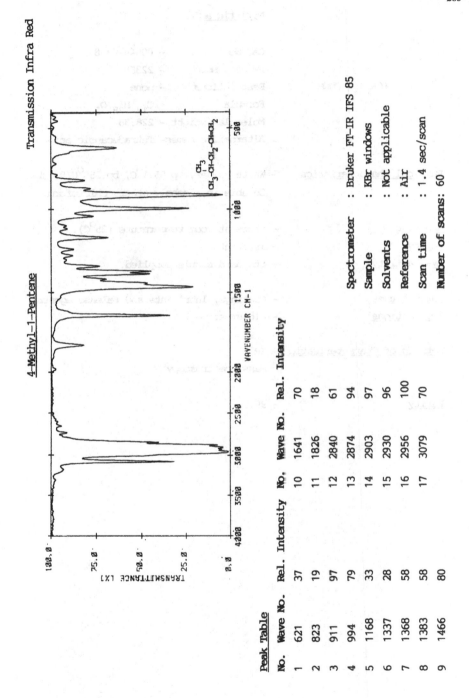

4-Methyl-1-Pentene

Transmission Infra Red

$CH_3-CH-CH_2-CH=CH_2$

Spectrometer	: Bruker FT-IR IFS 85
Sample	: KBr windows
Solvents	: Not applicable
Reference	: Air
Scan time	: 1.4 sec/scan
Number of scans:	60

Peak Table

No.	Wave No.	Rel. Intensity	No.	Wave No.	Rel. Intensity
1	621	37	10	1641	70
2	823	19	11	1826	18
3	911	97	12	2840	61
4	994	79	13	2874	94
5	1168	33	14	2903	97
6	1337	28	15	2930	96
7	1368	58	16	2956	100
8	1383	58	17	3079	70
9	1466	80			

Myristic acid

$CH_3-(CH_2)_{12}-COOH$

CAS No.	– 00544-63-8
PM Ref. No.	– 22350
Restrictions	– none
Formula	– $C_{14} H_{28} O_2$
Molecular weight	– 228.38
Alternative names	– Tetradecanoic acid

Physical Characteristics — White powder, mp 55.1^{O}C, bp 250°C/0.13bar. Soluble in alcohol, ether, chloroform.

Handling — Store at room temperature (25°C).

Safety — Irritant.

Availability — Standard sample supplied.

Current uses — Coatings, lubricants and release agents.

Applications — Closures.

Methods of Characterisation — IR

Mass Spectroscopy

Purity — 99.5%

Myristic acid

$$CH_3-(CH_2)_{12}-COOH$$

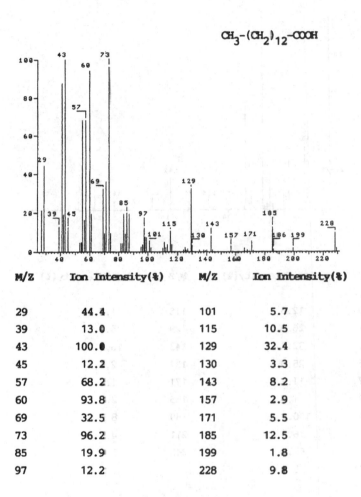

M/Z	Ion Intensity(%)	M/Z	Ion Intensity(%)
29	44.4	101	5.7
39	13.0	115	10.5
43	100.0	129	32.4
45	12.2	130	3.3
57	68.2	143	8.2
60	93.8	157	2.9
69	32.5	171	5.5
73	96.2	185	12.5
85	19.9	199	1.8
97	12.2	228	9.8

Spectrometer   :Finnigan Mat SSQ 70
Inlet System   :Capillary GC/MS
Source Temperature:150°C
Electron Energy   :70 eV
Scan Range   :25-400

Myristic acid, methyl ester

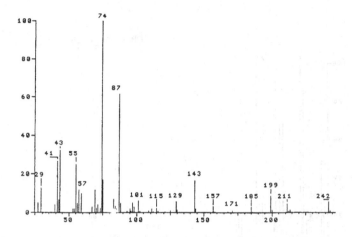

M/Z	Ion Intensity(%)	M/Z	Ion Intensity(%)
29	12.3	115	1.9
41	26.6	129	5.5
43	32.3	143	16.7
55	25.0	157	2.9
57	11.7	171	1.0
69	11.6	185	2.6
74	100.0	199	8.7
75	16.8	211	4.2
87	61.6	242	5.3
101	5.8		

Spectrometer   :Finnigan Mat SSQ 70
Inlet System   :Capillary GC/MS
Source Temperature:150°C
Electron Energy   :70 eV
Scan Range   :25-400

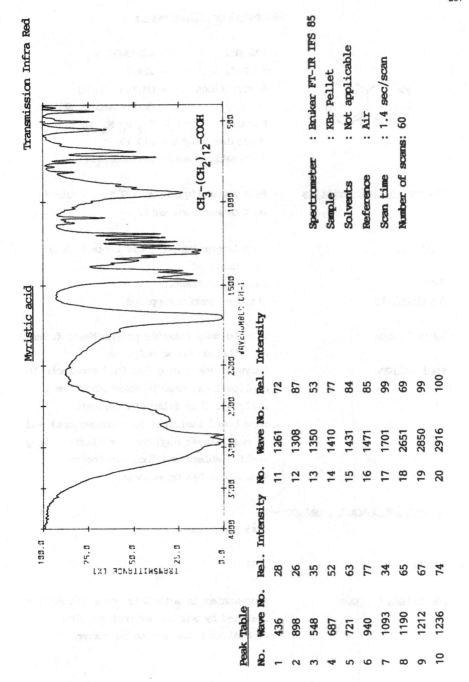

Myristic acid   Transmission Infra Red

$CH_3-(CH_2)_{12}-COOH$

Spectrometer	: Bruker FT-IR IFS 85
Sample	: KBr Pellet
Solvents	: Not applicable
Reference	: Air
Scan time	: 1.4 sec/scan
Number of scans:	60

Peak Table

No.	Wave No.	Rel. Intensity	No.	Wave No.	Rel. Intensity
1	436	28	11	1261	72
2	898	26	12	1308	87
3	548	35	13	1350	53
4	687	52	14	1410	77
5	721	63	15	1431	84
6	940	77	16	1471	85
7	1093	34	17	1701	99
8	1190	65	18	2651	69
9	1212	67	19	2850	99
10	1236	74	20	2916	100

1,5-Naphthalene diisocyanate

CAS No.	– 03173-72-6
PM Ref. No.	– 22420
Restrictions	– QM(T)= 1mg/kg (expressed as NCO).
Formula	– $C_{12} H_6 N_2 O_2$
Molecular weight	– 210.19
Alternative names	– Desmodur 15.

Physical Characteristics — Pale amber liquid, mp 127°C. Soluble in most organic solvents.

Handling — Room temperature (25°C). Protect from moisture.

Safety — Harmful/Irritant.

Availability — Standard sample supplied.

Current uses — Used to make flexible polyurethane foams and other urethane polymers.

Applications — Polyurethane tubing for food manufacturing applications. Used to make adhesives in seals for thin films, in polyester paperboard laminates (e.g susceptors) and in multi-layer high barrier plastics (e.g shelf stables) and 'boil-in-the-bag' laminates. Baking enamels.

Methods of Characterisation — IR
Mass Spectroscopy

Purity — >99%

Analytical methods — Isocyanates in materials and articles are analysed by solvent extraction with ethanol in toluene with concurrent urethane derivative formation, clean-up by liquid/liquid partition and solid phase cartridge chromatography and determination by capillary GC with nitrogen selective detection. Phenyl isocyanate and 1,4-butanediisocyanate are used as internal standards.

References   - Draft CEN Method (MAFF, FScL. Norwich, UK).

1,5-Naphthalene diisocyanate

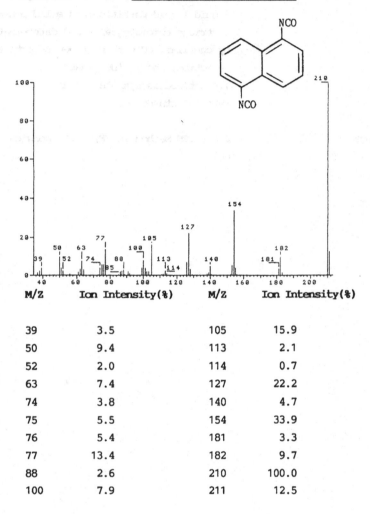

M/Z	Ion Intensity(%)	M/Z	Ion Intensity(%)
39	3.5	105	15.9
50	9.4	113	2.1
52	2.0	114	0.7
63	7.4	127	22.2
74	3.8	140	4.7
75	5.5	154	33.9
76	5.4	181	3.3
77	13.4	182	9.7
88	2.6	210	100.0
100	7.9	211	12.5

Spectrometer   :Finnigan Mat SSQ 70
Inlet System   :Capillary GC/MS
Source Temperature:150°C
Electron Energy   :70 eV
Scan Range   :25–400

1,5-Naphthalene diisocyanate   Transmission Infra Red

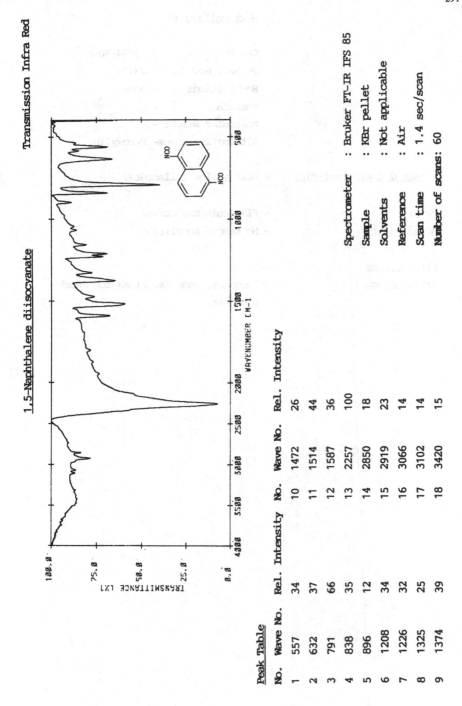

Spectrometer : Bruker FT-IR IFS 85
Sample   : KBr pellet
Solvents   : Not applicable
Reference   : Air
Scan time   : 1.4 sec/scan
Number of scans: 60

Peak Table

No.	Wave No.	Rel. Intensity	No.	Wave No.	Rel. Intensity
1	557	34	10	1472	26
2	632	37	11	1514	44
3	791	66	12	1587	36
4	838	35	13	2257	100
5	896	12	14	2850	18
6	1208	34	15	2919	23
7	1226	32	16	3066	14
8	1325	25	17	3102	14
9	1374	39	18	3420	15

Nitrocellulose

CAS No.   – 09004-70-0
PM Ref. No.   – 22450
Restrictions   – none
Formula   –
Molecular weight –
Alternative names– Pyroxylin.

Physical Characteristics   – Pale yellow fillaments.

Safety   – Flammable/Explosive.
Availability   – No sample supplied.

Current uses   –
Applications   – Lacquers, enamels. Films for snack
                     products.

1-Nonanol

$$CH_3-(CH_2)_7-CH_2OH$$

CAS No.	– 00143-08-8
PM Ref. No.	– 22480
Restrictions	– none
Formula	– $C_9 H_{20} O$
Molecular weight	– 144.26
Alternative names	– n-Nonyl alcohol.

Physical Characteristics
– Pale yellow liquid, mp –8 to –6°C, bp 215°C. Miscible with alcohol and ether.

Handling
– Store at room temperature (25°C).

Safety
– Combustible/Irritant.

Availability
– Standard sample supplied.

Current uses
– Polyester resins. Styrene resins. Vinyl chloride and polyolefin polymers.

Applications
– Low friction mouldings.

Methods of Characterisation
– IR
Mass Spectroscopy

Purity
– 99.6%

1-Nonanol

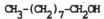

$CH_3-(CH_2)_7-CH_2OH$

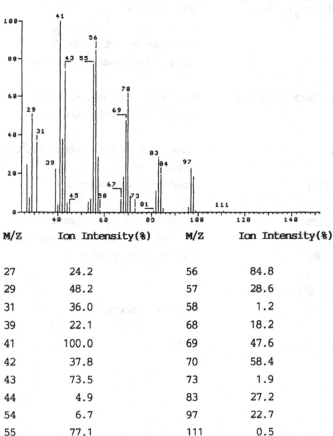

M/Z	Ion Intensity(%)	M/Z	Ion Intensity(%)
27	24.2	56	84.8
29	48.2	57	28.6
31	36.0	58	1.2
39	22.1	68	18.2
41	100.0	69	47.6
42	37.8	70	58.4
43	73.5	73	1.9
44	4.9	83	27.2
54	6.7	97	22.7
55	77.1	111	0.5

Spectrometer   :Finnigan Mat SSQ 70
Inlet System   :Capillary GC/MS
Source Temperature :150°C
Electron Energy   :70 eV
Scan Range   :25-400

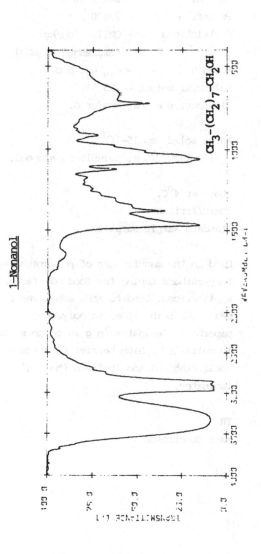

1-Nonanol

$CH_3-(CH_2)_7-CH_2OH$

Spectrometer	: Bruker FT-IR IFS 85
Sample	: KRS-5 windows
Solvents	: Not applicable
Reference	: Air
Scan time	: 1.4 sec/scan
Number of scans: 60	

Peak Table

No.	Wave No.	Rel. Intensity	No.	Wave No.	Rel. Intensity
1	662	43	6	1378	70
2	722	61	7	1466	91
3	912	30	8	2855	99
4	1058	91	9	2957	100
5	1121	41	10	3330	98

Octadecyl Isocyanate

$$CH_3-(CH_2)_{17}-NCO$$

CAS No.	— 00112–96–9
PM Ref. No.	— 22570
Restrictions	— QM(T)= 1mg/kg
	(expressed as NCO).
Formula	— $C_{19} H_{37} N O$
Molecular weight	— 295.51
Alternative names	— Mondur O.

Physical Characteristics — White solid, mp 15–16°C,
bp 172–1730°C/5mm. Soluble in alcohol.

Handling — Store at 4°C.

Safety — Toxic/Irritant.

Availability — Standard sample supplied.

Current uses — Used in the manufacture of polyurethanes.

Applications — Polyurethane tubing for food manufacturing applications. Used to make adhesives in seals for thin films, in polyester paperboard laminates (e.g susceptors) and in multi–layer high barrier plastics (e.g shelf stables) and 'boil–in–the–bag' laminates.

Methods of Characterisation — IR
Mass Spectroscopy

Purity — 98%

Analytical methods – Isocyanates in materials and articles are
analysed by solvent extraction with
ethanol in toluene with concurrent
urethane derivative formation, clean-up by
liquid/liquid partition and solid phase
cartridge chromatography and determination
by capillary GC with nitrogen selective
detection. Phenyl isocyanate and
1,4-butanediisocyanate are used as
internal standards.

References – Draft CEN Method (MAFF, FScL. Norwich,
UK).

298

Octadecyl isocyanate

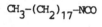

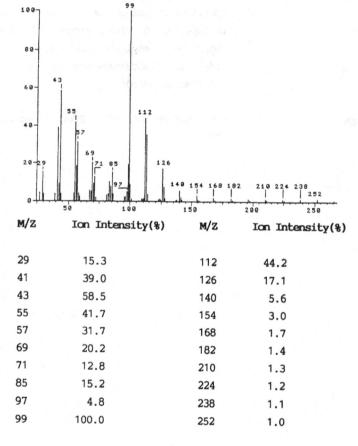

M/Z	Ion Intensity(%)	M/Z	Ion Intensity(%)
29	15.3	112	44.2
41	39.0	126	17.1
43	58.5	140	5.6
55	41.7	154	3.0
57	31.7	168	1.7
69	20.2	182	1.4
71	12.8	210	1.3
85	15.2	224	1.2
97	4.8	238	1.1
99	100.0	252	1.0

Spectrometer   :Finnigan Mat SSQ 70
Inlet System   :Capillary GC/MS
Source Temperature:150°C
Electron Energy   :70 eV
Scan Range   :25-400

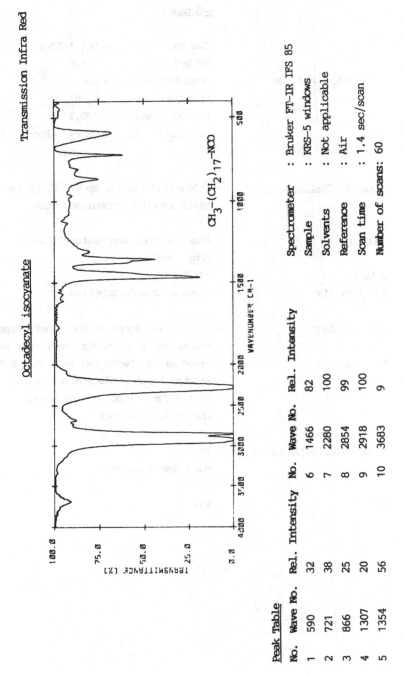

Octadecyl isocyanate   Transmission Infra Red

$CH_3-(CH_2)_{17}-NCO$

Spectrometer	:	Bruker FT-IR IFS 85
Sample	:	KRS-5 windows
Solvents	:	Not applicable
Reference	:	Air
Scan time	:	1.4 sec/scan
Number of scans:		60

Peak Table

No.	Wave No.	Rel. Intensity	No.	Wave No.	Rel. Intensity
1	590	32	6	1466	82
2	721	38	7	2280	100
3	866	25	8	2854	99
4	1307	20	9	2918	100
5	1354	56	10	3683	9

1-Octanol

$CH_3-(CH_2)_7-OH$

CAS No.	– 00111–87–5
PM Ref. No.	– 22600
Restrictions	– none
Formula	– $C_8 H_{18} O$
Molecular weight	– 130.23
Alternative names	– Capryl alcohol; octyl alcohol.

Physical Characteristics — Colourless liquid, mp $-17^{O}C$, bp 194–195°C. Miscible with alcohol and ether.

Handling — Store at room temperature (25°C). Anhydrous.

Safety — Irritant.

Availability — Standard sample supplied.

Current uses — A defoaming agent in the manufacture of cellophane base sheets. Polyester resins.

Applications — Bread wraps. Packaging for cakes and snacks. Fresh produce blisterpacks. Transparent window carton boxes. Coatings. Adhesives.

Methods of Characterisation — IR
Mass Spectroscopy

Purity — 99%

1-Octanol

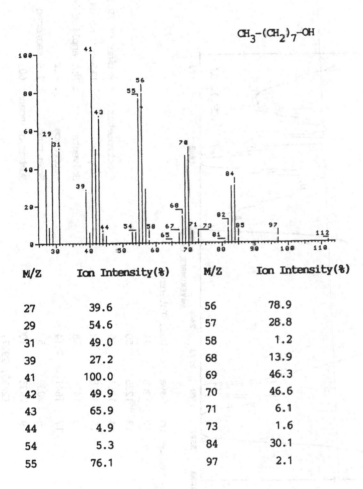

$$CH_3-(CH_2)_7-OH$$

M/Z	Ion Intensity(%)	M/Z	Ion Intensity(%)
27	39.6	56	78.9
29	54.6	57	28.8
31	49.0	58	1.2
39	27.2	68	13.9
41	100.0	69	46.3
42	49.9	70	46.6
43	65.9	71	6.1
44	4.9	73	1.6
54	5.3	84	30.1
55	76.1	97	2.1

Spectrometer   :Finnigan Mat SSQ 70
Inlet System   :Capillary GC/MS
Source Temperature:150°C
Electron Energy   :70 eV
Scan Range   :25-400

302

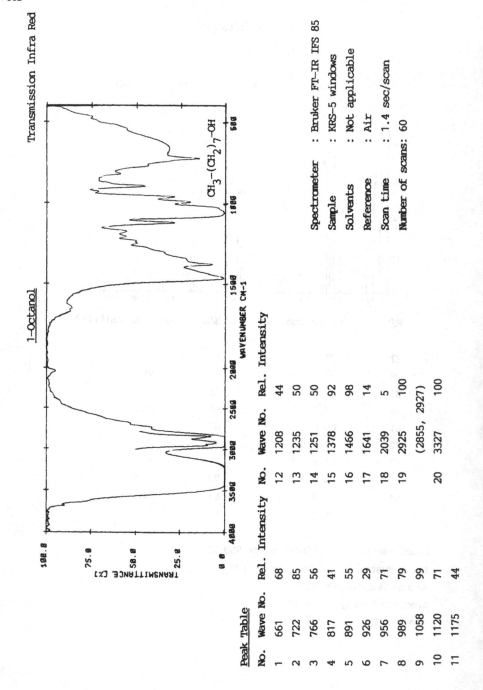

Transmission Infra Red

1-Octanol

$CH_3-(CH_2)_7-OH$

WAVENUMBER CM-1

TRANSMITTANCE [%]

Spectrometer	: Bruker FT-IR IFS 85
Sample	: KRS-5 windows
Solvents	: Not applicable
Reference	: Air
Scan time	: 1.4 sec/scan
Number of scans: 60	

Peak Table

No.	Wave No.	Rel. Intensity	No.	Wave No.	Rel. Intensity
1	661	68	12	1208	44
2	722	85	13	1235	50
3	766	56	14	1251	50
4	817	41	15	1378	92
5	891	55	16	1456	98
6	926	29	17	1641	14
7	956	71	18	2039	5
8	989	79	19	2925	100
9	1058	99		(2855, 2927)	
10	1120	71	20	3327	100
11	1175	44			

1-Octene

CAS No.	– 00111–66–0
PM Ref. No.	– 22660
Restrictions	– SML=15 mg/kg
Formula	– $C_8 H_{16}$
Molecular weight	– 112.22
Alternative names	– Caprylene

$CH_3-(CH_2)_5-CH=CH_2$

Physical Characteristics – Colourless liquid, mp –102°C, bp 121°C.

Handling – Store at room temperature (25°C).

Safety – Flammable/Irritant.

Availability – Standard sample supplied.

Current uses – A co-monomer for linear low density polyethylene.

Applications – Film for pre-packed fresh and frozen foods. Part of laminate with paper or aluminium. In bags, and blow moulded for storage containers.

Methods of Characterisation – IR
Mass Spectroscopy

Purity – 95%

Analytical methods – Headspace GC method with FID for 1-octene in polyethylene films (limit of detection = 0.01 mg/kg) and in olive oil and water

(0.01 mg/kg), aqueous ethanol and aqueous acetic acid (0.02 mg/kg) simulants.

References   – Method under development (Instituto Superiore di Sanita, Rome, I).
Food Addit. Contam., (1988) 5, 373–380.

1-Octene

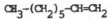

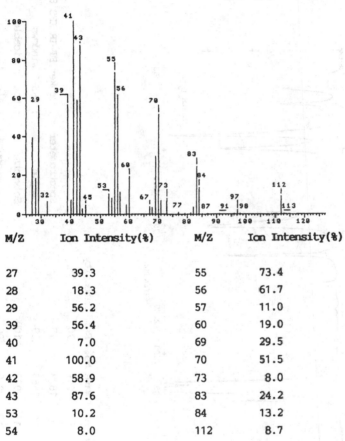

M/Z	Ion Intensity(%)	M/Z	Ion Intensity(%)
27	39.3	55	73.4
28	18.3	56	61.7
29	56.2	57	11.0
39	56.4	60	19.0
40	7.0	69	29.5
41	100.0	70	51.5
42	58.9	73	8.0
43	87.6	83	24.2
53	10.2	84	13.2
54	8.0	112	8.7

Spectrometer   :Finnigan Mat SSQ 70
Inlet System   :Capillary GC/MS
Source Temperature :150°C
Electron Energy   :70 eV
Scan Range   :25-400

306

1-Octene   Transmission Infra Red

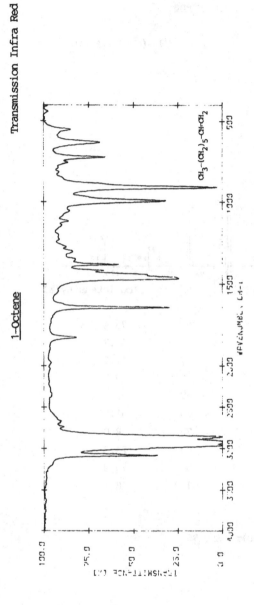

CH$_3$-(CH$_2$)$_5$-CH=CH$_2$

Spectrometer	: Bruker FT-IR IFS 85
Sample	: KRS-5 windows
Solvents	: Not applicable
Reference	: Air
Scan time	: 1.4 sec/scan
Number of scans: 60	

Peak Table

No.	Wave No.	Rel. Intensity	No.	Wave No.	Rel. Intensity
1	553	15	8	1641	70
2	634	32	9	1823	18
3	725	34	10	2856	99
4	909	96	11	2920	100
5	993	67	12	2959	99
6	1379	41	13	3079	63
7	1465	75			

Palmitic acid

$CH_3-(CH_2)_{14}-COOH$

CAS No.	— 00057–10–3
PM Ref. No.	— 22780
Restrictions	— none
Formula	— $C_{16} H_{32} O_2$
Molecular weight	— 256.43
Alternative names	— Hexadecanoic acid.

Physical Characteristics — White crystaline chips, mp 61–64°C.
Soluble in hot alcohol, ether, acetone and benzene.

Handling — Store at room temperature (25°C).

Safety — Irritant.

Availability — Standard sample supplied.

Current uses — Used to manufacture alkyd resins. Used in butadiene–styrene block co–polymers and methacrylic acid–styrene co–polymer mouldings. Used to make palmityl stearate.

Applications — Mould release agent. Scale inhibitor in the polymerisation of vinyl chloride on acrylic polymer latexes. Improves bonding in vulcanised rubber. Lubricant.

Methods of Characterisation — IR
Mass Spectroscopy

Purity — 99%

Palmitic acid

$CH_3-(CH_2)_{14}-COOH$

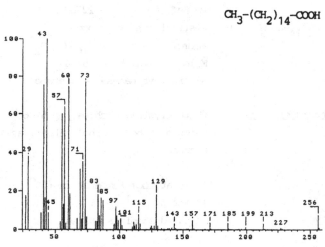

M/Z	Ion Intensity(%)	M/Z	Ion Intensity(%)
29	38.1	115	7.9
43	100.0	125	1.7
57	63.7	129	18.3
60	74.5	143	3.1
71	35.2	157	5.0
73	76.6	171	3.6
83	18.1	185	3.2
85	16.6	199	1.2
97	11.9	213	3.1
101	5.9	256	7.7

Spectrometer :Finnigan Mat SSQ 70
Inlet System :Capillary GC/MS
Source Temperature:150°C
Electron Energy :70 eV
Scan Range :25–400

Palmitic acid, methyl ester

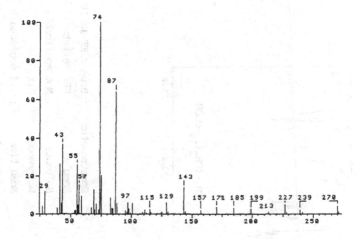

M/Z	Ion Intensity(%)	M/Z	Ion Intensity(%)
29	11.5	129	5.8
41	26.4	143	13.8
43	37.0	157	1.6
55	26.2	171	3.5
57	15.2	185	3.2
74	100.0	199	2.6
75	20.4	213	1.0
87	64.0	227	4.8
97	6.0	239	2.3
115	2.4	270	4.1

Spectrometer   :Finnigan Mat SSQ 70
Inlet System   :Capillary GC/MS
Source Temperature:150°C
Electron Energy   :70 eV
Scan Range   :25–400

310

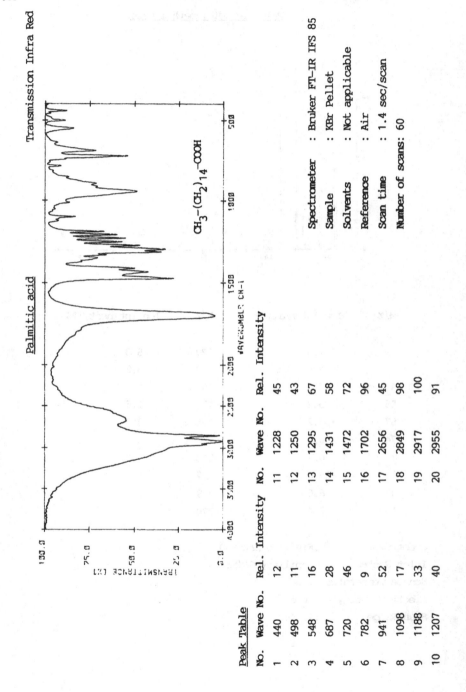

Palmitic acid

Transmission Infra Red

$CH_3-(CH_2)_{14}-COOH$

Spectrometer	: Bruker FT-IR IFS 85
Sample	: KBr Pellet
Solvents	: Not applicable
Reference	: Air
Scan time	: 1.4 sec/scan
Number of scans: 60	

Peak Table

No.	Wave No.	Rel. Intensity	No.	Wave No.	Rel. Intensity
1	440	12	11	1228	45
2	498	11	12	1250	43
3	548	16	13	1295	67
4	687	28	14	1431	58
5	720	46	15	1472	72
6	782	9	16	1702	96
7	941	52	17	2656	45
8	1098	17	18	2849	98
9	1188	33	19	2917	100
10	1207	40	20	2955	91

Pentaerythritol

$C(CH_2OH)_4$

CAS No.	– 0115–77–5
PM Ref. No.	– 22840
Restrictions	– none
Formula	– $C_5 H_{12} O_4$
Molecular weight	– 136.15
Alternative names	– Monopentaerythritol, tetramethylol methane.

Physical Characteristics – White crystalline powder, mp 269°C. Sublimes. Soluble in water, ethanol and glycerol.

Handling – Store at room temperature (25°C).

Safety – Irritant/Flammable.

Availability – Standard sample supplied.

Current uses – Cross-linking agent for polyester resin. Binder in printing inks. Imparts stability to PVC latexes. Silicones (hard) and waxes. Modifier for rosins. Co-polymer with adipic and stearic acid.

Applications – Coatings for food and beverage cans. Adhesives. Lubricant in the manufacture of rigid and semi-rigid polyvinyl chloride polymers.

Methods of Characterisation – IR
Mass Spectroscopy

Purity – 99%

Pentaerythritol

$$C(CH_2OH)_4$$

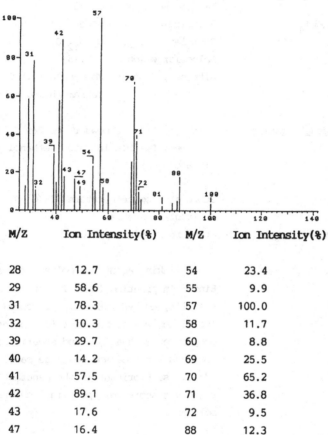

M/Z	Ion Intensity(%)	M/Z	Ion Intensity(%)
28	12.7	54	23.4
29	58.6	55	9.9
31	78.3	57	100.0
32	10.3	58	11.7
39	29.7	60	8.8
40	14.2	69	25.5
41	57.5	70	65.2
42	89.1	71	36.8
43	17.6	72	9.5
47	16.4	88	12.3

Spectrometer   :Finnigan Mat SSQ 70
Inlet System   :Capillary GC/MS
Source Temperature   :150°C
Electron Energy   :70 eV
Scan Range   :25–400

Transmission Infra Red

Pentaerythritol

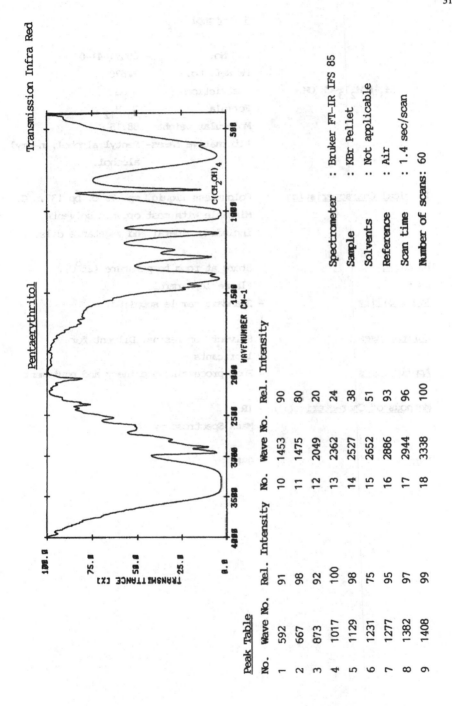

C(CH$_2$OH)$_4$

Spectrometer	: Bruker FT-IR IFS 85
Sample	: KBr Pellet
Solvents	: Not applicable
Reference	: Air
Scan time	: 1.4 sec/scan
Number of scans: 60	

Peak Table

No.	Wave No.	Rel. Intensity	No.	Wave No.	Rel. Intensity
1	592	91	10	1453	90
2	667	98	11	1475	80
3	873	92	12	2049	20
4	1017	100	13	2362	24
5	1129	98	14	2527	38
6	1231	75	15	2652	51
7	1277	95	16	2886	93
8	1382	97	17	2944	96
9	1408	99	18	3338	100

1-Pentanol

$CH_3-(CH_2)_3-CH_2OH$

CAS No.   – 00071-41-0
PM Ref. No.   – 22870
Restrictions   – none
Formula   – $C_5 H_{12} O$
Molecular weight – 88.15
Alternative names– Pentyl alcohol, n-amyl
                   alcohol.

Physical Characteristics   – Colourless liquid, mp -79°C, bp 137.5^{O}C.
                             Miscible with most organic solvents,
                             including mineral and vegetable oils.

Handling   – Store at room temperature (25°C).
Safety   – Flammable/Harmful.
Availability   – Standard sample supplied.

Current uses   – Solvent for resins. Diluent for
                    lubricants.
Applications   – Food processing machinery and equipment.

Methods of Characterisation – IR
                    Mass Spectroscopy

Purity   – 99%

1-Pentanol

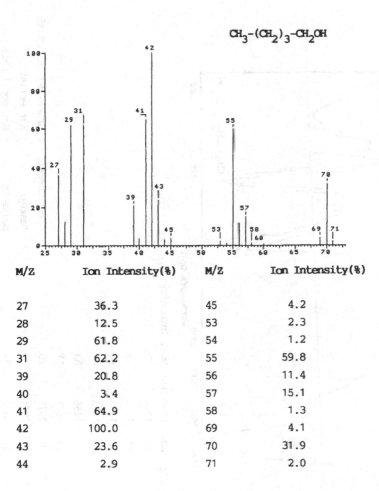

$CH_3-(CH_2)_3-CH_2OH$

M/Z	Ion Intensity(%)	M/Z	Ion Intensity(%)
27	36.3	45	4.2
28	12.5	53	2.3
29	61.8	54	1.2
31	62.2	55	59.8
39	20.8	56	11.4
40	3.4	57	15.1
41	64.9	58	1.3
42	100.0	69	4.1
43	23.6	70	31.9
44	2.9	71	2.0

Spectrometer   :Finnigan Mat SSQ 70
Inlet System   :Capillary GC/MS
Source Temperature:150°C
Electron Energy   :70 eV
Scan Range   :25–400

Transmission Infra Red

1-Pentanol

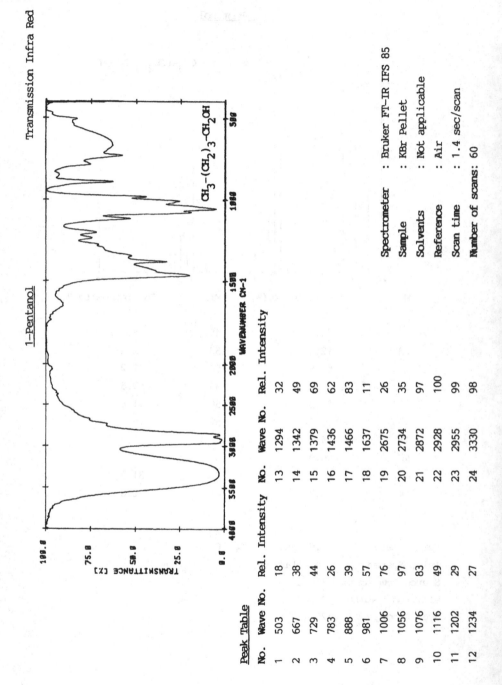

CH$_3$-(CH$_2$)$_3$-CH$_2$OH

Spectrometer	: Bruker FT-IR IFS 85
Sample	: KBr Pellet
Solvents	: Not applicable
Reference	: Air
Scan time	: 1.4 sec/scan
Number of scans: 60	

Peak Table

No.	Wave No.	Rel. Intensity	No.	Wave No.	Rel. Intensity
1	503	18	13	1294	32
2	667	38	14	1342	49
3	729	44	15	1379	69
4	783	26	16	1436	62
5	888	39	17	1466	83
6	981	57	18	1637	11
7	1006	76	19	2675	26
8	1056	97	20	2734	35
9	1076	83	21	2872	97
10	1116	49	22	2928	100
11	1202	29	23	2955	99
12	1234	27	24	3330	98

Phenol

OH

CAS No.	– 00108–95–2
PM Ref. No.	– 22960
Restrictions	– none
Formula	– $C_6 H_6 O$
Molecular weight	– 94.11
Alternative names	– Hydroxybenzene; carbolic acid.

Physical Characteristics – White crystals, mp 40–42°C, bp 182°C. Miscible with water. Soluble in alcohol, chloroform, ether and glycerol. Stabilised with hypophosphorous acid.

Handling – Store at room temperature (25°C). Protect from light and air.

Safety – Poison/Corrosive.

Availability – Standard sample supplied.

Current uses – Starting material for caprolactam (nylon 6) and bisphenol A. Co-polymer with formaldehyde. Blocking agent for isocyanates for polyurethane production.

Applications – Coatings for cans. 'Bakelite'. Screw cap closures. Lacquers. Coffee makers. Utensil and pan handles. Resins used to coat wine vats.

Methods of Characterisation – IR Mass Spectroscopy

Purity – 99%

318

Phenol

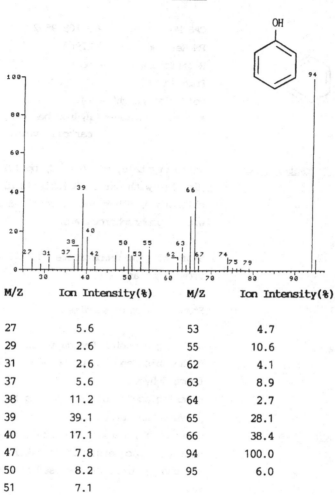

M/Z	Ion Intensity(%)	M/Z	Ion Intensity(%)
27	5.6	53	4.7
29	2.6	55	10.6
31	2.6	62	4.1
37	5.6	63	8.9
38	11.2	64	2.7
39	39.1	65	28.1
40	17.1	66	38.4
47	7.8	94	100.0
50	8.2	95	6.0
51	7.1		

Spectrometer   :Finnigan Mat SSQ 70
Inlet System   :Capillary GC/MS
Source Temperature:150°C
Electron Energy   :70 eV
Scan Range   :25–400

Phenol

Transmission Infra Red

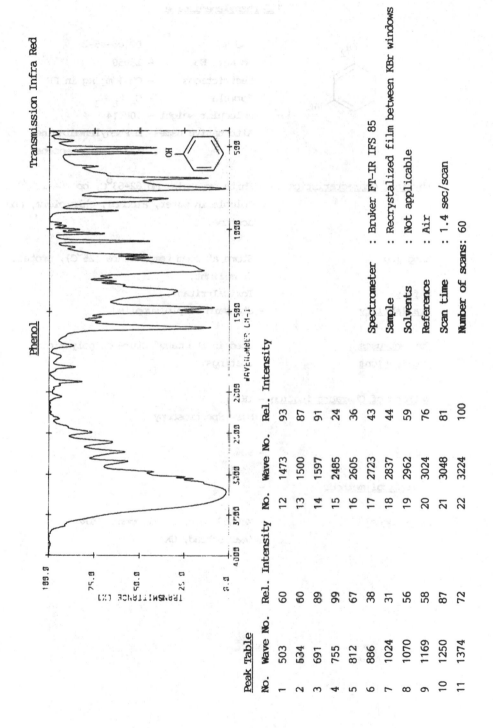

Spectrometer	: Bruker FT-IR IFS 85
Sample	: Recrystalized film between KBr windows
Solvents	: Not applicable
Reference	: Air
Scan time	: 1.4 sec/scan
Number of scans:	60

Peak Table

No.	Wave No.	Rel. Intensity	No.	Wave No.	Rel. Intensity
1	503	60	12	1473	93
2	534	60	13	1500	87
3	691	89	14	1597	91
4	755	99	15	2485	24
5	812	67	16	2605	36
6	886	38	17	2723	43
7	1024	31	18	2837	44
8	1070	56	19	2962	59
9	1169	58	20	3024	76
10	1250	87	21	3048	81
11	1374	72	22	3224	100

1,3-Phenylenediamine

NH$_2$

NH$_2$

CAS No.   – 00108–45–2
PM Ref. No.   – 23050
Restrictions   – QM=1 mg/kg in FP
Formula   – C$_6$ H$_8$ N$_2$
Molecular weight – 108.14
Alternative names– m-Phenylenediamine,
                   1,3–diaminobenzene

Physical Characteristics   – White crystals, mp 62–63°C, bp 284–287°C.
                              Soluble in water, alcohol, chloroform, and
                              acetone.

Handling   – Store at room temperature (25°C), protect
                              from light.

Safety   – Toxic/Irritant.

Availability   – Standard sample supplied.

Current uses   – Used in the manufacture of polyamides.

Applications   – Coatings.

Methods of Characterisation – IR
                              Mass Spectroscopy

Purity   – 98%

Analytical methods   –

References   – Method under development (PIRA,
                              Leatherhead, UK).

1,3-Phenylenediamine

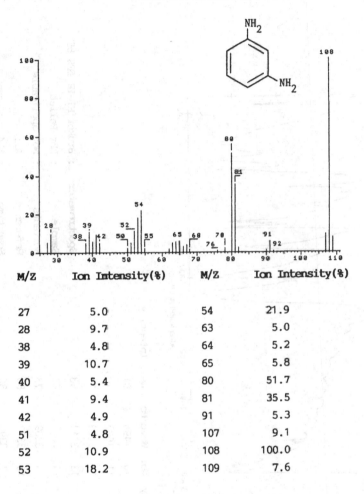

M/Z	Ion Intensity(%)	M/Z	Ion Intensity(%)
27	5.0	54	21.9
28	9.7	63	5.0
38	4.8	64	5.2
39	10.7	65	5.8
40	5.4	80	51.7
41	9.4	81	35.5
42	4.9	91	5.3
51	4.8	107	9.1
52	10.9	108	100.0
53	18.2	109	7.6

Spectrometer   :Finnigan Mat SSQ 70
Inlet System   :Capillary GC/MS
Source Temperature:150°C
Electron Energy   :70 eV
Scan Range   :25-400

1,3-Phenylenediamine   Transmission Infra Red

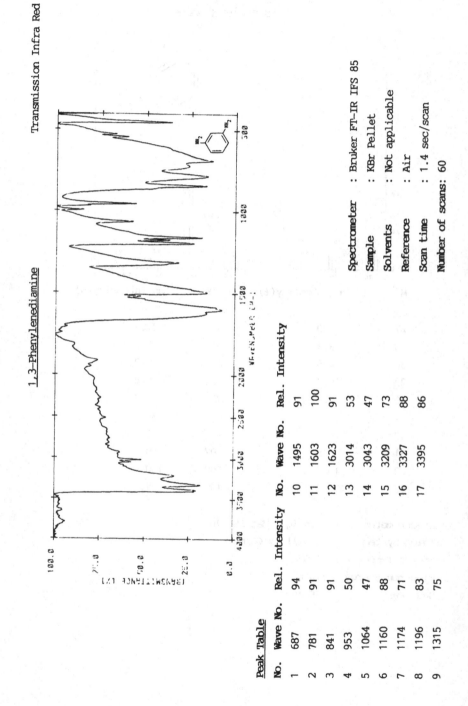

Spectrometer	: Bruker FT-IR IFS 85
Sample	: KBr Pellet
Solvents	: Not applicable
Reference	: Air
Scan time	: 1.4 sec/scan
Number of scans: 60	

Peak Table

No.	Wave No.	Rel. Intensity	No.	Wave No.	Rel. Intensity
1	687	94	10	1495	91
2	781	91	11	1603	100
3	841	91	12	1623	91
4	953	50	13	3014	53
5	1064	47	14	3043	47
6	1160	88	15	3209	73
7	1174	71	16	3327	88
8	1196	83	17	3395	86
9	1315	75			

Phenyl isocyanate

NCO

CAS No.	– 00103–71–9
PM Ref. No.	– 23125
Restrictions	– QM(T)= 1mg/kg in FP (expressed as NCO)
Formula	– $C_7 H_5 N O$
Molecular weight	– 119.12
Alternative names	– Isocyanatobenzene; Carbanil.

Physical Characteristics	– Liquid, mp –30°C, bp 158–168°C. Soluble in ether.
Handling	– Store at room temperature (25°C).
Safety	– Toxic/Corrosive.
Availability	– Standard sample supplied.
Current uses	– Used in the manufacture of polyurethanes.
Applications	– Polyurethane tubing for food manufacturing applications. Used to make adhesives in seals for thin films, in polyester paperboard laminates (e.g microwave susceptors) and in multi-layer high barrier plastics (e.g shelf stables) and 'boil-in-the-bag' laminates.
Methods of Characterisation	– IR Mass Spectroscopy
Purity	– 98%

Analytical methods — Isocyanates in materials and articles are analysed by solvent extraction with ethanol in toluene with concurrent urethane derivative formation, clean–up by liquid/liquid partition and solid phase cartridge chromatography and determination by capillary GC with nitrogen selective detection. Phenyl isocyanate and 1,4–butanediisocyanate are used as internal standards.

References — Draft CEN Method (MAFF, FScL. Norwich, UK).

Phenyl isocyanate

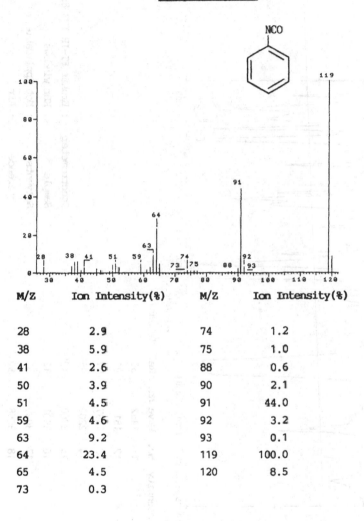

M/Z	Ion Intensity(%)	M/Z	Ion Intensity(%)
28	2.9	74	1.2
38	5.9	75	1.0
41	2.6	88	0.6
50	3.9	90	2.1
51	4.5	91	44.0
59	4.6	92	3.2
63	9.2	93	0.1
64	23.4	119	100.0
65	4.5	120	8.5
73	0.3		

Spectrometer   :Finnigan Mat SSQ 70
Inlet System   :Capillary GC/MS
Source Temperature:150°C
Electron Energy   :70 eV
Scan Range   :25-400

Transmission Infra Red

Phenyl isocyanate

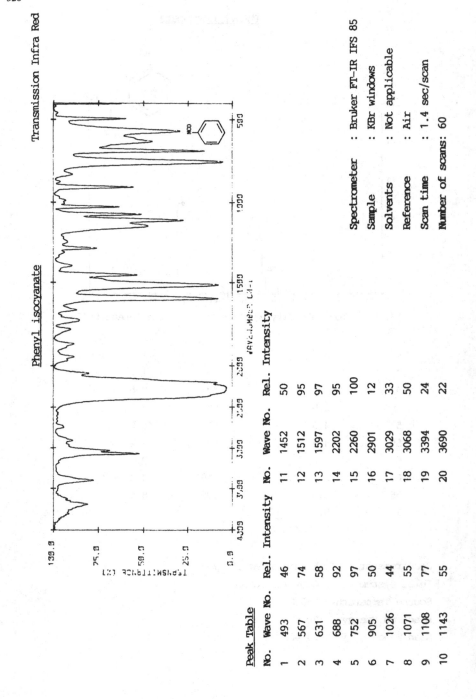

Spectrometer	: Bruker FT-IR IFS 85
Sample	: KBr windows
Solvents	: Not applicable
Reference	: Air
Scan time	: 1.4 sec/scan
Number of scans: 60	

Peak Table

No.	Wave No.	Rel. Intensity	No.	Wave No.	Rel. Intensity
1	493	46	11	1452	50
2	567	74	12	1512	95
3	631	58	13	1597	97
4	688	92	14	2202	95
5	752	97	15	2260	100
6	905	50	16	2901	12
7	1026	44	17	3029	33
8	1071	55	18	3068	50
9	1108	77	19	3394	24
10	1143	55	20	3690	22

Phosphoric acid

H_3PO_4

CAS No.   – 07664–38–2
PM Ref. No.   – 23170
Restrictions   – none
Formula   – $H_3\ O_4\ P$
Molecular weight – 98.0
Alternative names– Orthophosphoric acid.

Physical Characteristics   – White crystaline powder, bp $42^{O}C$.
                              Soluble in water.

Handling   – Hygroscopic.
Safety   – Corrosive/Irritant.
Availability   – Standard sample supplied.

Current uses   – Acid catalyst in the manufacture of
                    ethylene. Phosphoric acid esters of
                    polyester resins. Used to make polyethers
                    and polyurethanes.

Applications   – Coatings for aluminium.

Methods of Characterisation – IR

Purity   – 99%

Phosphoric acid   Transmission Infra Red

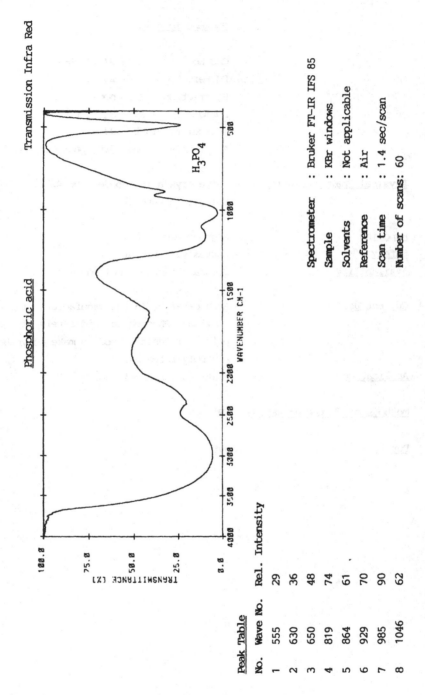

Spectrometer   : Bruker FT-IR IFS 85

Sample   : KBr windows

Solvents   : Not applicable

Reference   : Air

Scan time   : 1.4 sec/scan

Number of scans: 60

Peak Table

No.	Wave No.	Rel. Intensity
1	555	29
2	630	36
3	650	48
4	819	74
5	864	61
6	929	70
7	985	90
8	1046	62

Phthalic acid, diallyl ester

COOCH$_2$CH=CH$_2$

COOCH$_2$CH=CH$_2$

CAS No.	– 00131–17–9
PM Ref. No.	– 23230
Restrictions	– SML(T)= 0.01mg/kg (expressed as allyl).
Formula	– C$_{14}$ H$_{14}$ O$_4$
Molecular weight	– 246.27
Alternative names	– Diallyl phthalate.

Physical Characteristics   – Liquid, bp 165–167°C.

Handling   – Store at room temperature (25°C).

Safety   – Irritant/Harmful.

Availability   – Standard sample supplied.

Current uses   – Co-polymer with VdC and VC. Cross-linking agent for styrene based polyester resins. Used in acrylic co-polymers.

Applications   – Used for containers for fruit, vegetables, juices, wines and sauces. Adhesives.

Methods of Characterisation – IR

Mass Spectroscopy

Purity   – 97%

Analytical methods   –

References   –

Phthalic acid, diallyl ester

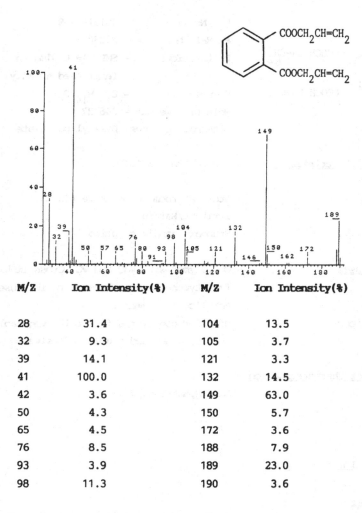

M/Z	Ion Intensity(%)	M/Z	Ion Intensity(%)
28	31.4	104	13.5
32	9.3	105	3.7
39	14.1	121	3.3
41	100.0	132	14.5
42	3.6	149	63.0
50	4.3	150	5.7
65	4.5	172	3.6
76	8.5	188	7.9
93	3.9	189	23.0
98	11.3	190	3.6

Spectrometer   :Finnigan Mat SSQ 70
Inlet System   :Capillary GC/MS
Source Temperature:150°C
Electron Energy   :70 eV
Scan Range   :25-400

331

Phthalic acid, diallyl ester
Transmission Infra Red

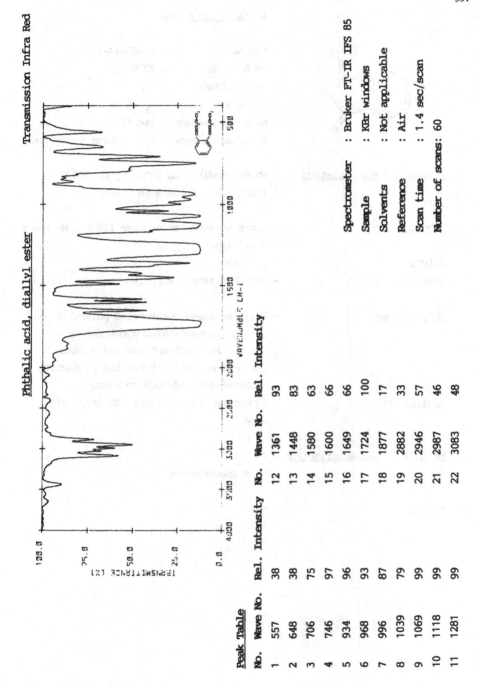

TRANSMITTANCE (%)
100.0
75.0
50.0
25.0
0.0

4000 3500 3000 2500 2000 1500 1000 500
WAVENUMBER CM-1

Spectrometer : Bruker FT-IR IFS 85
Sample   : KBr windows
Solvents   : Not applicable
Reference   : Air
Scan time   : 1.4 sec/scan
Number of scans: 60

Peak Table

No.	Wave No.	Rel. Intensity	No.	Wave No.	Rel. Intensity
1	557	38	12	1361	93
2	648	38	13	1448	83
3	706	75	14	1580	63
4	746	97	15	1600	66
5	934	96	16	1649	66
6	968	93	17	1724	100
7	996	87	18	1877	17
8	1039	79	19	2882	33
9	1069	99	20	2946	57
10	1118	99	21	2987	46
11	1281	99	22	3083	48

Phthalic anhydride

CAS No.	– 00085–44–9
PM Ref. No.	– 23380
Restrictions	– none
Formula	– $C_8 H_4 O_3$
Molecular weight	– 148.12
Alternative names	– 1,3–Isobenzofurandione.

Physical Characteristics — White needles, mp 130.8°C, bp 284°C (sublimes). Soluble in alcohol.

Handling — Store at room temperature (25°C). Moisture sensitive.

Safety — Irritant.

Availability — Standard sample supplied.

Current uses — Polyurethanes. Unsaturated polyester resins. Cross-linking agent for epoxy resins. As a solvent for alkyd resin production. Catalyst for the production of melamine/formaldehyde mouldings.

Applications — Adhesives. Can coatings. Table/picnic ware.

Methods of Characterisation — IR
Mass Spectroscopy

Purity — 99%

Phthalic anhydride

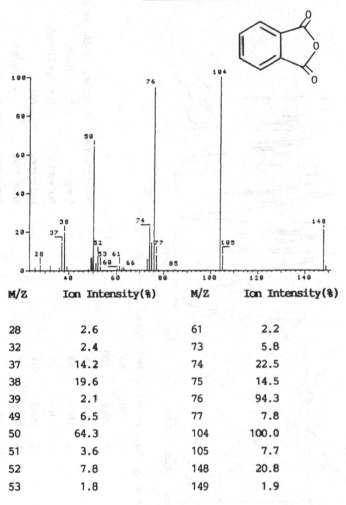

M/Z	Ion Intensity(%)	M/Z	Ion Intensity(%)
28	2.6	61	2.2
32	2.4	73	5.8
37	14.2	74	22.5
38	19.6	75	14.5
39	2.1	76	94.3
49	6.5	77	7.8
50	64.3	104	100.0
51	3.6	105	7.7
52	7.8	148	20.8
53	1.8	149	1.9

Spectrometer   :Finnigan Mat SSQ 70
Inlet System   :Capillary GC/MS
Source Temperature:150°C
Electron Energy   :70 eV
Scan Range   :25–400

334

Phthalic anhydride   Transmission Infra Red

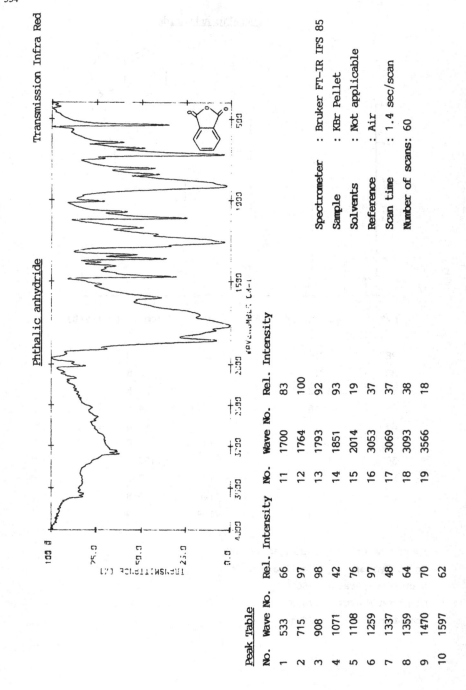

Spectrometer : Bruker FT-IR IFS 85
Sample : KBr Pellet
Solvents : Not applicable
Reference : Air
Scan time : 1.4 sec/scan
Number of scans: 60

Peak Table

No.	Wave No.	Rel. Intensity	No.	Wave No.	Rel. Intensity
1	533	66	11	1700	83
2	715	97	12	1764	100
3	908	98	13	1793	92
4	1071	42	14	1851	93
5	1108	76	15	2014	19
6	1259	97	16	3053	37
7	1337	48	17	3069	37
8	1359	64	18	3093	38
9	1470	70	19	3566	18
10	1597	62			

alpha-Pinene

CAS No.	– 00080-56-8
PM Ref. No.	– 23470
Restrictions	– none
Formula	– $C_{10}H_{16}$
Molecular weight	– 136.24
Alternative names	– 2,6,6-Trimethylbicyclo [3.1.1]hept-2-ene; 2-Pinene.

Physical Characteristics – Liquid, mp -50°C, bp 155-156°C. Insoluble in water, soluble in alcohol, chloroform, ether and glacial acetic acid.

Safety – Harmful/Flammable/Irritant.

Availability – No standard sample with this CAS No.

Current uses – Terpene resins. Co-polymer with maleic acid.

Applications – Terpene resin blends used to form laminates with good printability and moisture resistance. Coatings.

336

CAS No.	– 00127–91–3
PM Ref. No.	– 23500
Restrictions	– none
Formula	– $C_{10} H_{16}$
Molecular weight	– 136.23
Alternative names	– Nopinene.

Physical Characteristics — Colourless liquid, bp 164–166°C.
Soluble in alcohol, ether, benzene, and chloroform.

Handling — Store at room temperature (25°C).

Safety — Flammable/Irritant.

Availability — Standard sample supplied.

Current uses — Terpene resins. Co-polymer with maleic acid.

Applications — Terpene resin blends are used to form laminates with printability and moisture resistance. Coatings.

Methods of Characterisation — IR
Mass Spectroscopy

Purity — 99%

beta-Pinene

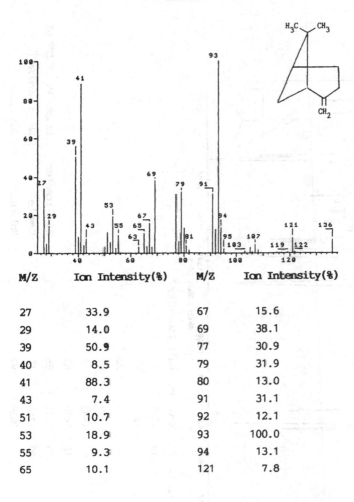

M/Z	Ion Intensity(%)	M/Z	Ion Intensity(%)
27	33.9	67	15.6
29	14.0	69	38.1
39	50.9	77	30.9
40	8.5	79	31.9
41	88.3	80	13.0
43	7.4	91	31.1
51	10.7	92	12.1
53	18.9	93	100.0
55	9.3	94	13.1
65	10.1	121	7.8

Spectrometer   :Finnigan Mat SSQ 70
Inlet System   :Capillary GC/MS
Source Temperature:150°C
Electron Energy   :70 eV
Scan Range   :25-400

338

Transmission Infra Red

beta-Pinene

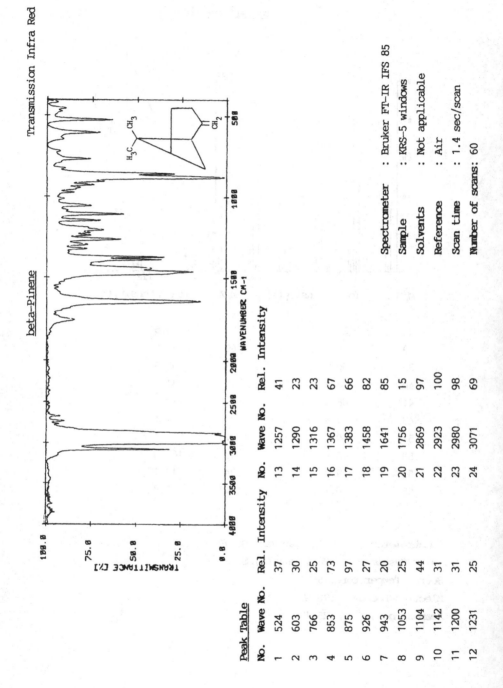

WAVENUMBER CM-1

Spectrometer : Bruker FT-IR IFS 85
Sample : KRS-5 windows
Solvents : Not applicable
Reference : Air
Scan time : 1.4 sec/scan
Number of scans: 60

Peak Table

No.	Wave No.	Rel. Intensity	No.	Wave No.	Rel. Intensity
1	524	37	13	1257	41
2	603	30	14	1290	23
3	766	25	15	1316	23
4	853	73	16	1367	67
5	875	97	17	1383	66
6	926	27	18	1458	82
7	943	20	19	1641	85
8	1053	25	20	1756	15
9	1104	44	21	2869	97
10	1142	31	22	2923	100
11	1200	31	23	2980	98
12	1231	25	24	3071	69

Polyethylene glycol

$HOCH_2CH_2-[OCH_2CH_2OH]_n$

CAS No.	– 25322–68–3
PM Ref. No.	– 23590
Restrictions	– SML(T)= 30mg/kg
Formula	–
Molecular weight	– variable
Alternative names	– Macrogol.

Physical Characteristics – Viscous liquid or white solid, depending on the value of n. Soluble in water.

Safety –
Availability – No sample supplied.

Current uses – Polyester resins.
Applications – Re-use and single-use trays for pre-cooked, frozen and chilled meals. Carbonated beverage bottles. Storage tanks.

Polypropyleneglycol

$$OH-[C_3H_8-O-]_n-H$$

CAS No.	– 25322-69-4
PM Ref. No.	– 23650
Restrictions	– none
Formula	– variable
Molecular weight	– >400.
Alternative names–	

Physical Characteristics – Colourless liquid.

Handling   – Store at room temperature (25°C).

Safety   – Irritant.

Availability   – Standard sample supplied.

Current uses   – Co-polymer used to make polyurethane
             resins, polyvinylidene chloride, and
             acrylic and methacrylic resins.

Applications   – Sterilizable containers. Snap fit
             closures. Electric kettles. Plates.
             Adhesives. Coatings.

Methods of Characterisation – IR

Purity   – 99%

341

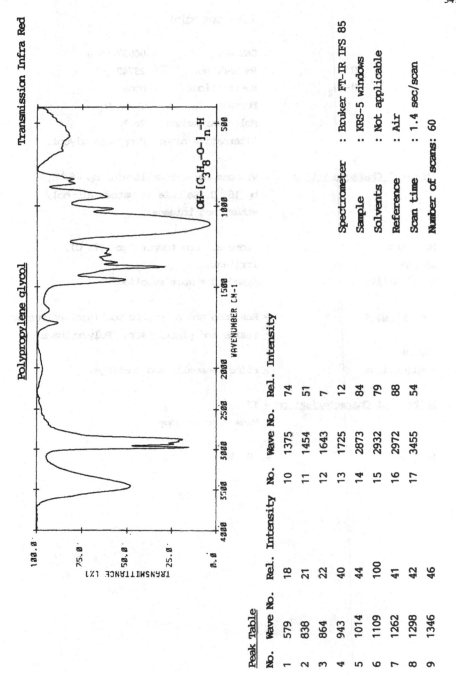

Polypropylene glycol

Transmission Infra Red

$OH-[C_3H_8-O-]_n-H$

Spectrometer : Bruker FT-IR IFS 85
Sample : KRS-5 windows
Solvents : Not applicable
Reference : Air
Scan time : 1.4 sec/scan
Number of scans: 60

Peak Table

No.	Wave No.	Rel. Intensity	No.	Wave No.	Rel. Intensity
1	579	18	10	1375	74
2	838	21	11	1454	51
3	864	22	12	1643	7
4	943	40	13	1725	12
5	1014	44	14	2873	84
6	1109	100	15	2932	79
7	1262	41	16	2972	88
8	1298	42	17	3455	54
9	1346	46			

1,2-Propanediol

OH OH

$CH_3-CH-CH_2$

CAS No.	– 00057-55-6
PM Ref. No.	– 23740
Restrictions	– none
Formula	– $C_3 H_8 O_2$
Molecular weight	– 76.10
Alternative names	– Propylene glycol.

Physical Characteristics
- Viscous colourless liquid, mp -60°C, bp 187°C. Soluble in water, alcohol, ether, and toluene.

Handling
- Store at room temperature (25°C).

Safety
- Irritant.

Availability
- Standard sample supplied.

Current uses
- Resinous and polymeric coatings. Polyester resins and plasticisers. Polyurethane resins.

Applications
- Films, enamels, and coatings.

Methods of Characterisation
- IR
Mass Spectroscopy

Purity
– 99.5%

1,2-Propanediol

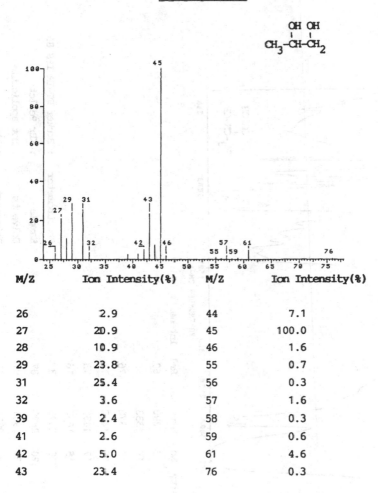

M/Z	Ion Intensity(%)	M/Z	Ion Intensity(%)
26	2.9	44	7.1
27	20.9	45	100.0
28	10.9	46	1.6
29	23.8	55	0.7
31	25.4	56	0.3
32	3.6	57	1.6
39	2.4	58	0.3
41	2.6	59	0.6
42	5.0	61	4.6
43	23.4	76	0.3

Spectrometer   :Finnigan Mat SSQ 70
Inlet System   :Capillary GC/MS
Source Temperature:150°C
Electron Energy   :70 eV
Scan Range   :25-400

Transmission Infra Red

1,2-Propanediol

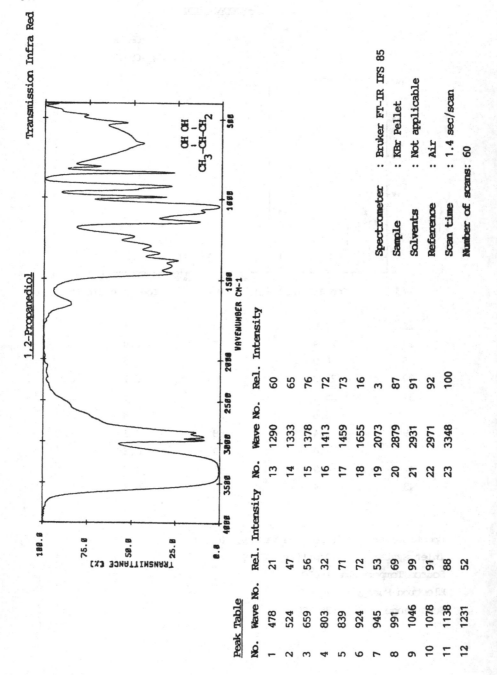

$CH_3-CH-CH_2$
   | |
   OH   OH

Spectrometer	:	Bruker FT-IR IFS 85
Sample	:	KBr Pellet
Solvents	:	Not applicable
Reference	:	Air
Scan time	:	1.4 sec/scan
Number of scans: 60		

Peak Table

No.	Wave No.	Rel. Intensity	No.	Wave No.	Rel. Intensity
1	478	21	13	1290	60
2	524	47	14	1333	65
3	659	56	15	1378	76
4	803	32	16	1413	72
5	839	71	17	1459	73
6	924	72	18	1655	16
7	945	53	19	2073	3
8	991	69	20	2879	87
9	1046	99	21	2931	91
10	1078	91	22	2971	92
11	1138	88	23	3348	100
12	1231	52			

1-Propanol

$CH_3-CH_2-CH_2OH$

CAS No.   – 00071-23-8
PM Ref. No.   – 23800
Restrictions   – none
Formula   – $C_3 H_8 O$
Molecular weight – 60.10
Alternative names – n-Propyl alcohol.

Physical Characteristics – Colourless liquid, mp -127°C, bp 97.2°C.
Soluble in water, alcohol, ether, acetone
and benzene.

Handling   – Store at room temperature (25°C).
Anhydrous.

Safety   – Flammable/Irritant.

Availability   – Standard sample supplied.

Current uses   – Solvent for flexographic printing inks.

Applications   – Printing on polyolefin and polyamide
films.

Methods of Characterisation – IR
Mass Spectroscopy

Purity   – 99.5%

1-Propanol

$CH_3-CH_2-CH_2OH$

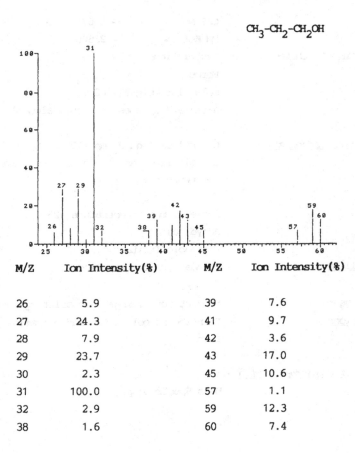

M/Z	Ion Intensity(%)	M/Z	Ion Intensity(%)
26	5.9	39	7.6
27	24.3	41	9.7
28	7.9	42	3.6
29	23.7	43	17.0
30	2.3	45	10.6
31	100.0	57	1.1
32	2.9	59	12.3
38	1.6	60	7.4

Spectrometer   :Finnigan Mat SSQ 70
Inlet System   :Capillary GC/MS
Source Temperature:150°C
Electron Energy   :70 eV
Scan Range   :25-400

347

Transmission Infra Red

1-Propanol

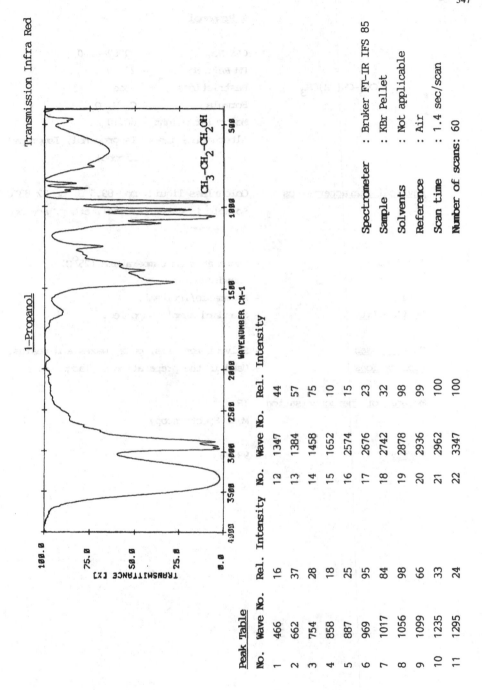

$CH_3-CH_2-CH_2OH$

Spectrometer	: Bruker FT-IR IFS 85
Sample	: KBr Pellet
Solvents	: Not applicable
Reference	: Air
Scan time	: 1.4 sec/scan
Number of scans: 60

Peak Table

No.	Wave No.	Rel. Intensity	No.	Wave No.	Rel. Intensity
1	466	16	12	1347	44
2	662	37	13	1384	57
3	754	28	14	1458	75
4	858	18	15	1652	10
5	887	25	16	2574	15
6	969	95	17	2676	23
7	1017	84	18	2742	32
8	1056	98	19	2878	98
9	1099	66	20	2936	99
10	1235	33	21	2962	100
11	1295	24	22	3347	100

2-Propanol

$CH_3-CH(OH)CH_3$

CAS No.   – 00067-63-0
PM Ref. No.   – 23830
Restrictions   – none
Formula   – $C_3 H_8 O$
Molecular weight   – 60.10
Alternative names–   Isopropanol, Isopropyl
                     alcohol.

Physical Characteristics   – Colourless liquid, mp -89.5°C, bp 82-83°C.
                             Soluble in water, alcohol, ether, acetone
                             and benzene.

Handling   – Store at room temperature (25°C).
                       Anhydrous.
Safety   – Flammable/Irritant.
Availability   – Standard sample supplied.

Current uses   – Solvent for oils, gums, waxes and resins.
Applications   – Used in the preparation of lacquers.

Methods of Characterisation – IR
                       Mass Spectroscopy

Purity   – 99.7%

2-Propanol

$CH_3-CH(OH)CH_3$

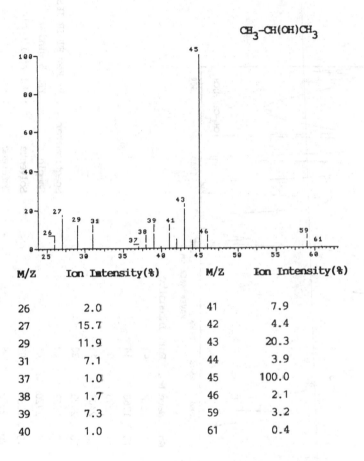

M/Z	Ion Intensity(%)	M/Z	Ion Intensity(%)
26	2.0	41	7.9
27	15.7	42	4.4
29	11.9	43	20.3
31	7.1	44	3.9
37	1.0	45	100.0
38	1.7	46	2.1
39	7.3	59	3.2
40	1.0	61	0.4

Spectrometer   :Finnigan Mat SSQ 70
Inlet System   :Capillary GC/MS
Source Temperature:150°C
Electron Energy   :70 eV
Scan Range   :25-400

Transmission Infra Red

2-Propanol

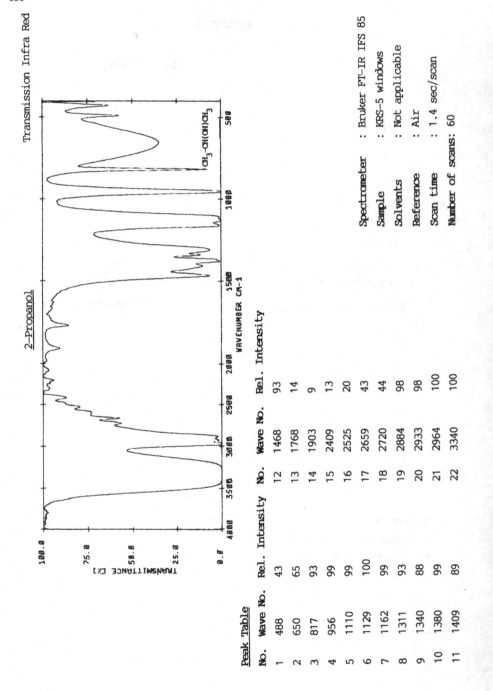

CH₃—CH(OH)CH₃

WAVENUMBER CM-1

TRANSMITTANCE [%]

Spectrometer	: Bruker FT-IR IFS 85
Sample	: KRS-5 windows
Solvents	: Not applicable
Reference	: Air
Scan time	: 1.4 sec/scan
Number of scans: 60	

Peak Table

No.	Wave No.	Rel. Intensity	No.	Wave No.	Rel. Intensity
1	488	43	12	1468	93
2	650	65	13	1768	14
3	817	93	14	1903	9
4	956	99	15	2409	13
5	1110	99	16	2525	20
6	1129	100	17	2659	43
7	1162	99	18	2720	44
8	1311	93	19	2884	98
9	1340	88	20	2933	98
10	1380	99	21	2964	100
11	1409	89	22	3340	100

Propionaldehyde

$$CH_3-CH_2-CHO$$

CAS No.   – 00123–38–6
PM Ref. No.   – 23860
Restrictions   – none
Formula   – $C_3 H_6 O$
Molecular weight – 58.08
Alternative names– Propanal, propionic
                    aldehyde.

Physical Characteristics   – Liquid, mp $-81^{o}C$, bp $47.5-49^{o}C$.
                                Suffocating odour. Soluble in water.

Safety   – Toxic/Irritant.
Availability   – No sample supplied.

Current uses   – Starting substance in the synthesis of
                                methacrylic acid, methyl methacrylate,
                                methacrylonitrile and acrylonitrile.
Applications   – Coatings. Film and sheeting. Rigid and
                                semi-rigid containers. Kitchen appliances.
                                Refigerator fittings.

Propionic acid

CH$_3$-CH$_2$-COOH

CAS No.	– 00079-09-4
PM Ref. No.	– 23890
Restrictions	– none
Formula	– C$_3$ H$_6$ O$_2$
Molecular weight	– 74.08
Alternative names	– Propanoic acid.

Physical Characteristics – Colourless liquid, mp –23°C, bp 141°C. Soluble in water, alcohol and ether.

Handling – Store at room temperature (25°C).

Safety – Toxic/Corrosive.

Availability – Standard sample supplied.

Current uses – Starting substance in the synthesis of methacrylic acid and its methyl ester.

Applications – Coatings. Film and sheeting. Rigid and semi-rigid containers.

Methods of Characterisation – IR
Mass Spectroscopy

Purity – 99%

Propionic acid

CH_3-CH_2-COOH

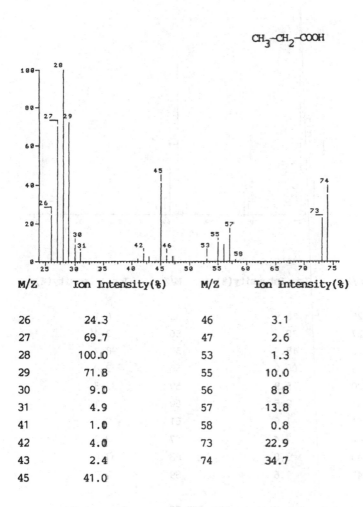

M/Z	Ion Intensity(%)	M/Z	Ion Intensity(%)
26	24.3	46	3.1
27	69.7	47	2.6
28	100.0	53	1.3
29	71.8	55	10.0
30	9.0	56	8.8
31	4.9	57	13.8
41	1.0	58	0.8
42	4.0	73	22.9
43	2.4	74	34.7
45	41.0		

Spectrometer	:Finnigan Mat SSQ 70
Inlet System	:Capillary GC/MS
Source Temperature	:150°C
Electron Energy	:70 eV
Scan Range	:25-400

Propionic acid, methyl ester

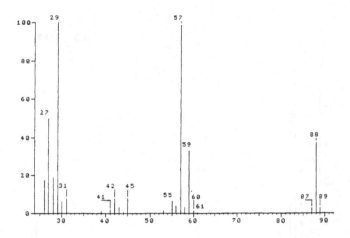

M/Z	Ion Intensity(%)	M/Z	Ion Intensity(%)
26	16.9	55	6.1
27	50.0	56	3.4
28	18.7	57	98.7
29	100.0	58	3.0
30	5.9	59	33.3
31	7.5	60	1.1
41	1.6	61	0.3
42	7.4	87	2.5
43	3.0	88	37.6
45	7.6	89	2.7

Spectrometer   :Finnigan Mat SSQ 70
Inlet System   :Capillary GC/MS
Source Temperature:150°C
Electron Energy   :70 eV
Scan Range   :25–400

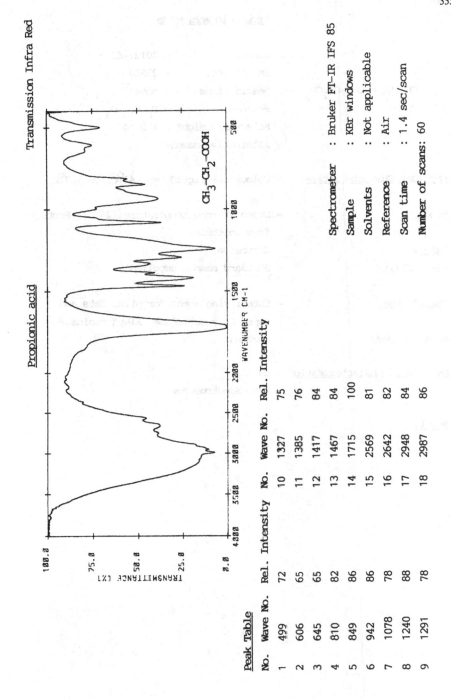

Propionic acid

Transmission Infra Red

CH_3-CH_2-COOH

Spectrometer	: Bruker FT-IR IFS 85
Sample	: KBr windows
Solvents	: Not applicable
Reference	: Air
Scan time	: 1.4 sec/scan
Number of scans:	60

Peak Table

No.	Wave No.	Rel. Intensity	No.	Wave No.	Rel. Intensity
1	499	72	10	1327	75
2	606	65	11	1385	76
3	645	65	12	1417	84
4	810	82	13	1467	84
5	849	86	14	1715	100
6	942	86	15	2569	81
7	1078	78	16	2642	82
8	1240	88	17	2948	84
9	1291	78	18	2987	86

Propionic anhydride

$$CH_3CH_2-\overset{O}{\overset{\|}{C}}-O-\overset{O}{\overset{\|}{C}}-CH_2CH_3$$

CAS No.	– 00123-62-6
PM Ref. No.	– 23950
Restrictions	– none
Formula	– $C_6 H_{10} O_3$
Molecular weight	– 130.14
Alternative names	–

Physical Characteristics   – Colourless liquid, mp –43°C, bp 167°C.

Handling   – Store at room temperature (25°C). Protect from moisture.

Safety   – Corrosive.

Availability   – Standard sample supplied.

Current uses   – Esterifying agent for oils, fats and especially cellulose. Alkyd resins.

Applications   – Coatings.

Methods of Characterisation – IR
Mass Spectroscopy

Purity   –

Propionic anhydride

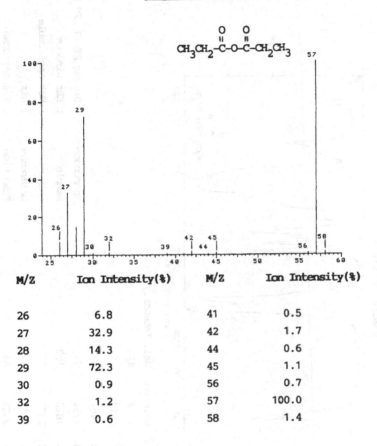

M/Z	Ion Intensity(%)	M/Z	Ion Intensity(%)
26	6.8	41	0.5
27	32.9	42	1.7
28	14.3	44	0.6
29	72.3	45	1.1
30	0.9	56	0.7
32	1.2	57	100.0
39	0.6	58	1.4

Spectrometer   :Finnigan Mat SSQ 70
Inlet System   :Capillary GC/MS
Source Temperature:150°C
Electron Energy   :70 eV
Scan Range   :25-400

Propionic anhydride

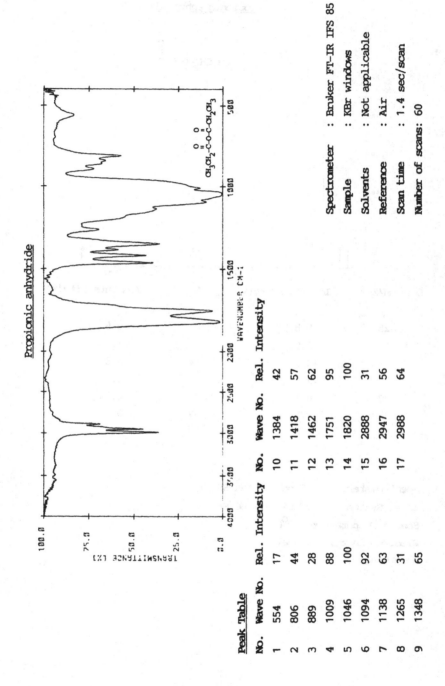

$$CH_3CH_2-C-O-C-CH_2CH_3$$

Spectrometer	: Bruker FT-IR IFS 85
Sample	: KBr windows
Solvents	: Not applicable
Reference	: Air
Scan time	: 1.4 sec/scan
Number of scans: 60	

Peak Table

No.	Wave No.	Rel. Intensity	No.	Wave No.	Rel. Intensity
1	554	17	10	1384	42
2	806	44	11	1418	57
3	889	28	12	1462	62
4	1009	88	13	1751	95
5	1046	100	14	1820	100
6	1094	92	15	2888	31
7	1138	63	16	2947	56
8	1265	31	17	2988	64
9	1348	65			

Propylene

CH₃-CH=CH₂

CAS No.	- 00115-07-1
PM Ref. No.	- 23980
Restrictions	- none
Formula	- $C_3 H_6$
Molecular weight	- 42.08
Alternative names	- 1-Propene, methylethylene.

Physical Characteristics - Colourless gas, mp -185°C, bp -48°C.

Safety - Flammable.

Availability - No sample supplied.

Current uses - Polypropylene polymers. As a co-monomer with vinyl chloride. Co-extruded with EVOH co-polymer. Extruded fibres, and textiles.

Applications - Used in the manufacture of injection moulded containers for food storage and display/transport, drinking straws. Moulds for cheese production, kettles. As a biaxially orientated or tubular quenched film used as an alternative to regenerated cellulose film (cellophane) for wrapping snack foods, biscuits, coffee and fruit. 'Thin-wall' margarine tubs. Bottles. Advanced barrier property packages for oven ready meals, & containers. Net for fruit packaging & woven textile bulk containers for granular products, powders and vegetables.

360

Propylene oxide

CH₃-CH - CH₂ (with O bridge)

CAS No.   – 00075-56-9
PM Ref. No.   – 24010
Restrictions   – QM= 1mg/kg
Formula   – C_3H_6O
Molecular weight – 58.08
Alternative names– Methyl oxirane.

Physical Characteristics – Colourless liquid, mp –112.1°C, bp 34.2°C. Soluble in toluene and water, miscible with alcohol and ether.

Handling – Refrigerate (4°C).

Safety – Cancer suspect agent/Flammable.

Availability – Standard sample supplied, as a solution in toluene at a concentration of 10mg/ml.

Current uses – Used to make polyether polyols. A co-polymer with starch. For removal of residual catalyst from crude polyolefins. Polyester resins.

Applications – Paper coatings and adhesives. Coatings for wine vats.

Methods of Characterisation – Mass Spectroscopy

Purity – 99%

Analytical methods – Solid sample heated in sealed vial and headspace sample analysed by GC with flame ionisation detection.

References – Method under development (Packforsk, Stockholm, S).
Zh. Anal. Khim., 1984, 39, 100-105.

Propylene oxide

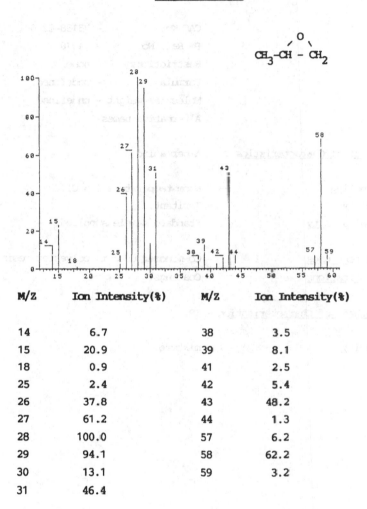

M/Z	Ion Intensity(%)	M/Z	Ion Intensity(%)
14	6.7	38	3.5
15	20.9	39	8.1
18	0.9	41	2.5
25	2.4	42	5.4
26	37.8	43	48.2
27	61.2	44	1.3
28	100.0	57	6.2
29	94.1	58	62.2
30	13.1	59	3.2
31	46.4		

Spectrometer   :Finnigan Mat SSQ 70
Inlet System   :Capillary GC/MS
Source Temperature:150°C
Electron Energy   :70 eV
Scan Range   :25-400

Resin acids and rosin acids

CAS No.   − 73138−82−6
PM Ref. No.   − 24070
Restrictions   − none
Formula   − undefined.
Molecular weight − undefined.
Alternative names−

Physical Characteristics   − Amber solid.

Handling   − Room temperature (25°C).
Safety   − Irritant.
Availability   − Standard sample supplied.

Current uses   − Co-monomer used in polyester resins.
Applications   − Coatings.

Methods of Characterisation − IR

Purity   − mixture.

Resin acids and Rosin acids   Transmission Infra Red

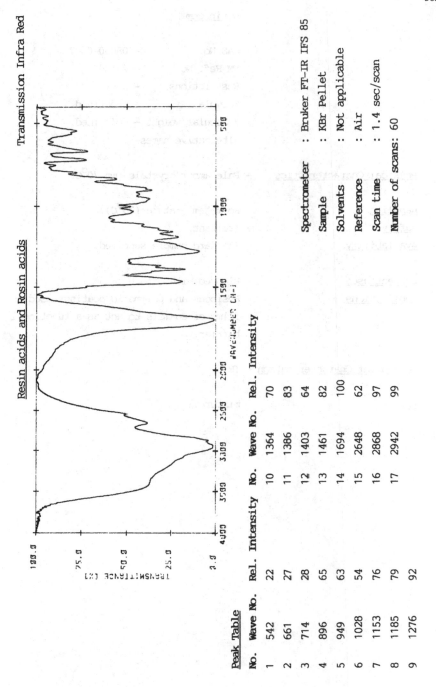

Spectrometer	: Bruker FT-IR IFS 85
Sample	: KBr Pellet
Solvents	: Not applicable
Reference	: Air
Scan time	: 1.4 sec/scan
Number of scans: 60	

Peak Table

No.	Wave No.	Rel. Intensity	No.	Wave No.	Rel. Intensity
1	542	22	10	1364	70
2	661	27	11	1386	83
3	714	28	12	1403	64
4	896	65	13	1461	82
5	949	63	14	1694	100
6	1028	54	15	2648	62
7	1153	76	16	2868	97
8	1185	79	17	2942	99
9	1276	92			

Rosin gum

CAS No.	– 08050–09–7
PM Ref. No.	– 24130
Restrictions	– none
Formula	– undefined.
Molecular weight	– undefined.
Alternative names–	

Physical Characteristics — Pale amber crystals, mp 70°C.

Handling — Room temperature (25°C).

Safety — Irritant.

Availability — Standard sample supplied.

Current uses — Blended with oils.

Applications — Resinous and polymeric coatings used as films or enamels to act as a functional barrier.

Methods of Characterisation — IR

Purity — mixture.

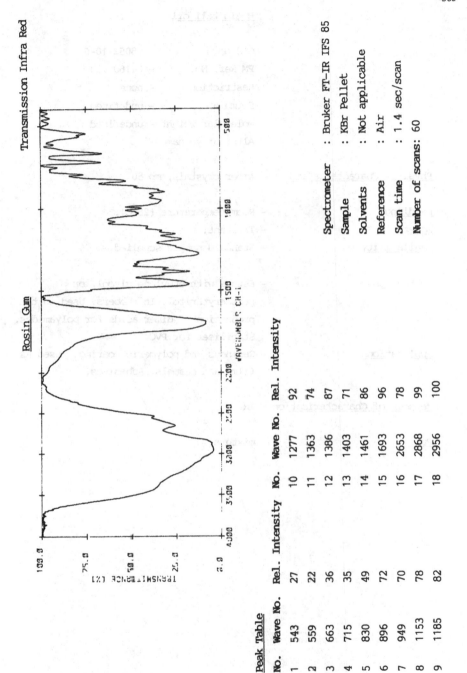

Rosin Gum   Transmission Infra Red

Spectrometer	: Bruker FT-IR IFS 85
Sample	: KBr Pellet
Solvents	: Not applicable
Reference	: Air
Scan time	: 1.4 sec/scan
Number of scans: 60	

Peak Table

No.	Wave No.	Rel. Intensity	No.	Wave No.	Rel. Intensity
1	543	27	10	1277	92
2	559	22	11	1363	74
3	663	36	12	1386	87
4	715	35	13	1403	71
5	830	49	14	1461	86
6	896	72	15	1693	96
7	949	70	16	2653	78
8	1153	78	17	2868	99
9	1185	82	18	2956	100

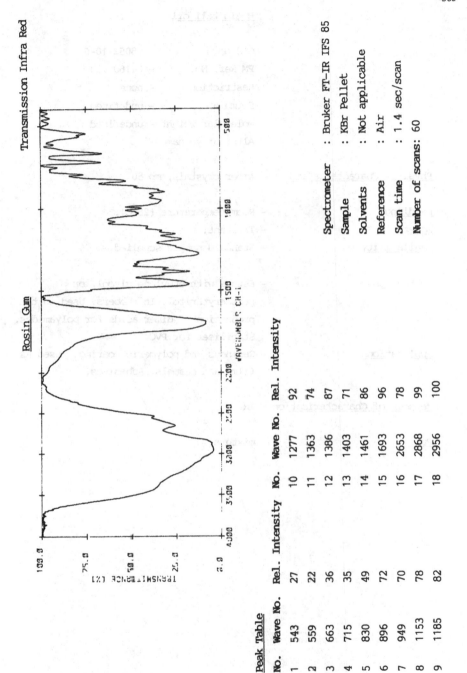

Rosin tall oil

CAS No.	– 08052-10-6
PM Ref. No.	– 24160
Restrictions	– none
Formula	– undefined
Molecular weight	– undefined
Alternative names	–

Physical Characteristics — Amber crystals, mp 80°C.

Handling — Room temperature (25°C).

Safety — Irritant.

Availability — Standard sample supplied.

Current uses — Esters with ethylene glycol, or pentaerythritol. In rubbers. Used in the production of dimer acids for polyamides. Stabiliser for PVC.

Applications — Resinous and polymeric coatings, used as films and enamels. Adhesives.

Methods of Characterisation — IR

Purity — mixture.

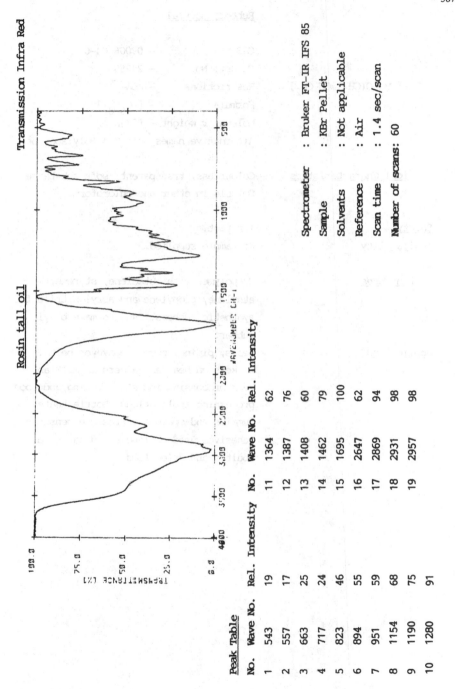

367

Rosin tall oil   Transmission Infra Red

Spectrometer : Bruker FT-IR IFS 85
Sample   : KBr Pellet
Solvents   : Not applicable
Reference   : Air
Scan time   : 1.4 sec/scan

Number of scans: 60

Peak Table

No.	Wave No.	Rel. Intensity	No.	Wave No.	Rel. Intensity
1	543	19	11	1364	62
2	557	17	12	1387	76
3	663	25	13	1408	60
4	717	24	14	1462	79
5	823	46	15	1695	100
6	894	55	16	2647	62
7	951	59	17	2869	94
8	1154	68	18	2931	98
9	1190	75	19	2957	98
10	1280	91			

Rubber, natural

$-[CH_2-C(CH_3)=CH-CH_2]-n$

CAS No. $-$ 09006-04-6
PM Ref. No. $-$ 24250
Restrictions $-$ none
Formula   · $-$ [$-C_5H_8-$]n
Molecular weight $-$ [68]n
Alternative names$-$ cis 1,4-Polyisoprene.

Physical Characteristics $-$ Colourless, transparent, soft substance. Soluble in ether and chloroform.

Safety $-$ Irritant.

Availability $-$ No sample supplied.

Current uses $-$ Co-polymer with butadiene, styrene, ethylene, propylene and acrylonitrile for synthetic rubbers. Used to make butyl rubber.

Applications $-$ Seals, piping, pumps, conveyer belts. Gaskets on heat exchangers in milk and beer processing industry. Tubing and food processing applications. Bottle teats, dummies and spatulas. Pressure sensitive adhesives used on labels and tapes for poultry and dried foods.

Sebacic acid

$$HOOC-(CH_2)_8-COOH$$

CAS No.	– 00111-20-6
PM Ref. No.	– 24280
Restrictions	– none
Formula	– $C_{10}H_{18}O_4$
Molecular weight	– 202.25
Alternative names	– Decanedioic acid.

Physical Characteristics — White powder, mp 134.5^{O}C, bp 294.5^{O}C/0.13 bar. Soluble in alcohol and ether.

Handling — Store at room temperature (25°C).

Safety — Irritant.

Availability — Standard sample supplied.

Current uses — Nucleating agent and clarity improver for polyolefins. Blends with Vinylidene Chloride co-polymers for high impact strength and low gas permeability. Binder for printing inks. Laminate with PET, and PS. Laminated with colour developers. Blends with polyamides for adhesives. Nylon 6/10 Poly(hexamethylene) sebacamide. Film laminates with polyolefins and methacrylic acid.

Applications — Bottles. Heat sealable films. Microwave bags. 'Heating time' indicators on food packaging. Adhesive for can lids. Gas barrier laminates.

Methods of Characterisation — IR
Mass Spectroscopy

Purity — 99%

Sebacic acid

$HOOC-(CH_2)_8-COOH$

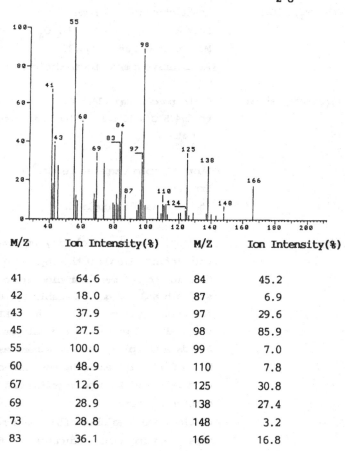

M/Z	Ion Intensity(%)	M/Z	Ion Intensity(%)
41	64.6	84	45.2
42	18.0	87	6.9
43	37.9	97	29.6
45	27.5	98	85.9
55	100.0	99	7.0
60	48.9	110	7.8
67	12.6	125	30.8
69	28.9	138	27.4
73	28.8	148	3.2
83	36.1	166	16.8

Spectrometer   :Finnigan Mat SSQ 70
Inlet System   :Capillary GC/MS
Source Temperature :150°C
Electron Energy   :70 eV
Scan Range   :25-400

Sebacic acid, dimethyl ester

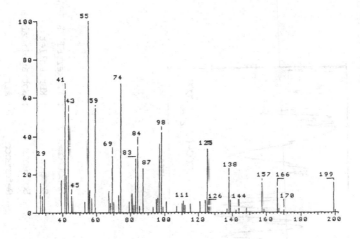

M/Z	Ion Intensity(%)	M/Z	Ion Intensity(%)
29	27.9	87	23.0
41	63.7	98	42.4
43	52.0	111	5.8
45	8.5	125	33.1
55	100.0	126	2.7
59	54.7	138	18.3
69	29.6	144	2.1
74	67.1	157	15.2
83	28.3	166	12.4
84	34.6	199	15.4

Spectrometer   :Finnigan Mat SSQ 70
Inlet System   :Capillary GC/MS
Source Temperature:150°C
Electron Energy   :70 eV
Scan Range   :25-400

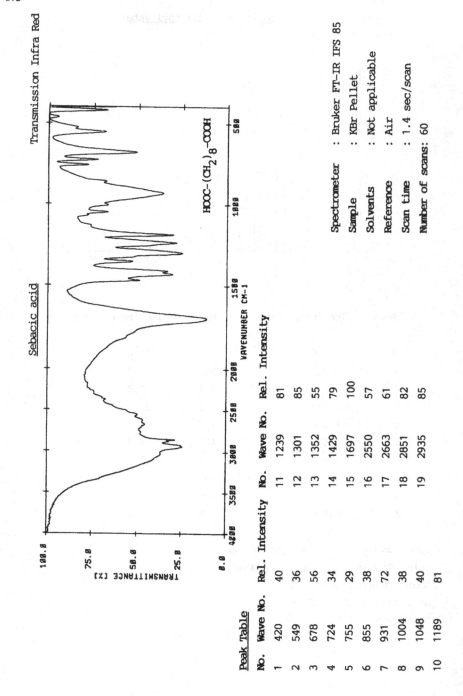

Sebacic acid

Transmission Infra Red

HOOC-(CH$_2$)$_8$-COOH

Spectrometer	: Bruker FT-IR IFS 85
Sample	: KBr Pellet
Solvents	: Not applicable
Reference	: Air
Scan time	: 1.4 sec/scan
Number of scans: 60	

Peak Table

No.	Wave No.	Rel. Intensity	No.	Wave No.	Rel. Intensity
1	420	40	11	1239	81
2	549	36	12	1301	85
3	678	56	13	1352	55
4	724	34	14	1429	79
5	755	29	15	1697	100
6	855	38	16	2550	57
7	931	72	17	2663	61
8	1004	38	18	2851	82
9	1048	40	19	2935	85
10	1189	81			

Sorbitol

CAS No.	– 00050–70–4
PM Ref. No.	– 24490
Restrictions	– none
Formula	– $C_6 H_{14} O_6$
Molecular weight	– 182.17
Alternative names	– D–glucitol.

$$HO-CH_2-CHOH-CHOH-CHOH$$
$$|$$
$$CHOH$$
$$|$$
$$CH_2OH$$

Physical Characteristics — White powder, mp 93–97°C. Soluble in water, acetone, and acetic acid.

Handling — Store at room temperature (25°C).

Safety — Irritant.

Availability — Standard sample supplied.

Current uses — Used to make cross–linked polyester resins. Starting substance in the synthesis of 1,2–propanediol. Polyurethane ionomer blends.

Applications — Finishes for polyester fibres for good adhesion to rubber. Coatings. Aqueous adhesives for plastic film laminates.

Methods of Characterisation — IR

Purity — 99%

Transmission Infra Red

Sorbitol

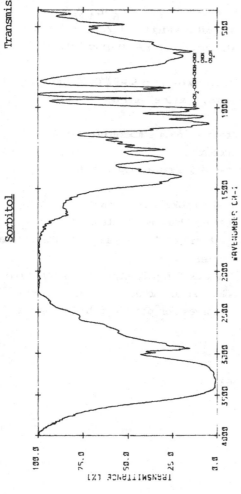

Spectrometer : Bruker FT-IR IFS 85
Sample : KBr Pellet
Solvents : Not applicable
Reference : Air
Scan time : 1.4 sec/scan
Number of scans: 60

Peak Table

No.	Wave No.	Rel. Intensity	No.	Wave No.	Rel. Intensity
1	481	48	9	1049	93
2	644	87	10	1095	96
3	675	82	11	1255	71
4	872	74	12	1309	71
5	887	71	13	1419	81
6	938	52	14	2932	85
7	1001	91	15	3282	100
8	1016	85			

Soybean oil

CAS No.   – 08001–22–7
PM Ref. No.   – 24520
Restrictions   – none
Formula   – undefined.
Molecular weight – undefined.
Alternative names–

Physical Characteristics   – Amber liquid.

Handling   – Room temperature (25°C).
Safety   – Irritant.
Availability   – Standard sample supplied.

Current uses   – Lubricant. Epoxidised soybean oil used as
                              a stabiliser for PVC.

Applications   – Adhesives. Coatings.

Methods of Characterisation – IR

Purity   – Natural product, composition variable.

376

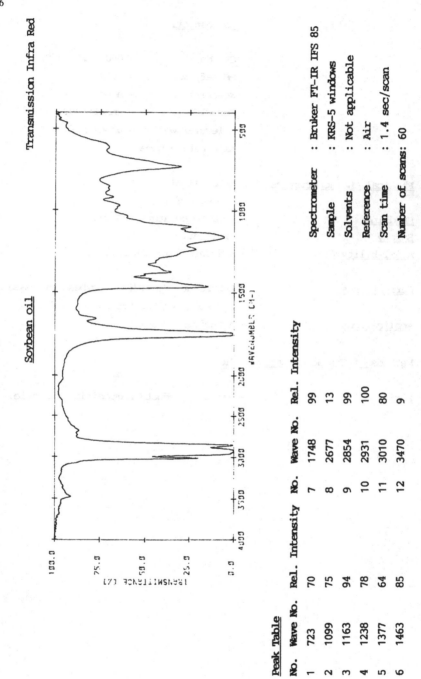

Soybean oil
Transmission Infra Red

Spectrometer : Bruker FT-IR IFS 85
Sample : KRS-5 windows
Solvents : Not applicable
Reference : Air
Scan time : 1.4 sec/scan
Number of scans: 60

Peak Table

No.	Wave No.	Rel. Intensity	No.	Wave No.	Rel. Intensity
1	723	70	7	1748	99
2	1099	75	8	2677	13
3	1163	94	9	2854	99
4	1238	78	10	2931	100
5	1377	64	11	3010	80
6	1463	85	12	3470	9

Stearic acid

$$CH_3-(CH_2)_{16}-COOH$$

CAS No.	– 00057-11-4
PM Ref. No.	– 24550
Restrictions	– none
Formula	– $C_{18} H_{36} O_2$
Molecular weight	– 284.49
Alternative names	– Octadecanoic acid,

Physical Characteristics   – White powder, mp 67-69°C, bp 361°C.
Soluble in ether, acetone and benzene.

Handling   – Store at room temperature (25°C).

Safety   – Irritant.

Availability   – Standard sample supplied.

Current uses   – Used to make polyvinyl stearate and
stearyl palmitate. Used to manufacture
alkyd resins. Co-polymer with adipic acid.

Applications   – Resinous coatings. Surface lubricant.

Methods of Characterisation – IR
Mass Spectroscopy

Purity   – 99%

Stearic acid

$$CH_3-(CH_2)_{16}-COOH$$

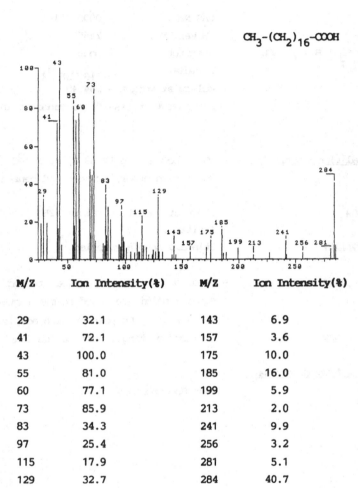

M/Z	Ion Intensity(%)	M/Z	Ion Intensity(%)
29	32.1	143	6.9
41	72.1	157	3.6
43	100.0	175	10.0
55	81.0	185	16.0
60	77.1	199	5.9
73	85.9	213	2.0
83	34.3	241	9.9
97	25.4	256	3.2
115	17.9	281	5.1
129	32.7	284	40.7

Spectrometer   :Finnigan Mat SSQ 70
Inlet System   :Capillary GC/MS
Source Temperature :$150^{\circ}C$
Electron Energy   :70 eV
Scan Range   :25-400

Stearic acid, methyl ester

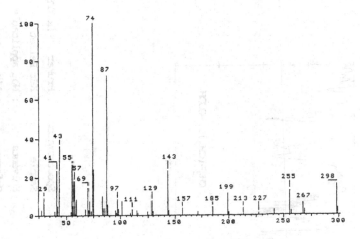

M/Z	Ion Intensity(%)	M/Z	Ion Intensity(%)
29	9.3	129	7.6
41	23.8	143	23.4
43	36.2	157	3.4
55	26.8	185	4.3
57	18.2	199	11.3
69	14.5	213	3.5
74	100.0	227	1.1
87	72.2	255	12.7
97	8.3	267	6.2
111	3.1	298	16.0

Spectrometer   :Finnigan Mat SSQ 70
Inlet System   :Capillary GC/MS
Source Temperature:150°C
Electron Energy   :70 eV
Scan Range   :25-400

380

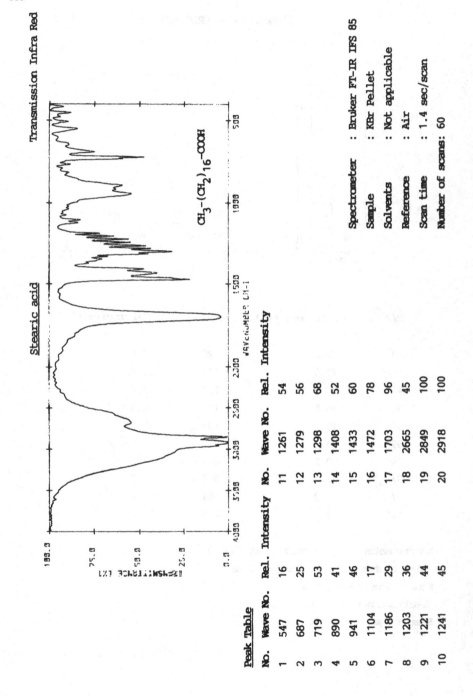

Transmission Infra Red

Stearic acid

$CH_3-(CH_2)_{16}-COOH$

Spectrometer	: Bruker FT-IR IFS 85
Sample	: KBr Pellet
Solvents	: Not applicable
Reference	: Air
Scan time	: 1.4 sec/scan
Number of scans: 60	

Peak Table

No.	Wave No.	Rel. Intensity	No.	Wave No.	Rel. Intensity
1	547	16	11	1261	54
2	687	25	12	1279	56
3	719	53	13	1298	68
4	890	41	14	1408	52
5	941	46	15	1433	60
6	1104	17	16	1472	78
7	1186	29	17	1703	96
8	1203	36	18	2665	45
9	1221	44	19	2849	100
10	1241	45	20	2918	100

Styrene

CAS No.	— 00100—42—5
PM Ref. No.	— 24610
Restrictions	— SML=0.6 mg/kg (under review).
Formula	— $C_8 H_8$
Molecular weight	— 104.15
Alternative names	— Cinnamol, Vinyl Benzene.

Physical Characteristics
— Clear liquid, mp $-30.6^{\circ}C$, bp $145^{\circ}C$, soluble in alcohol, acetone and benzene. Inhibited with 10—15 mg/kg, tertiary butyl catechol.

Handling
— Refrigerate, store below $18^{\circ}C$.

Safety
— Suspect carcinogen/Toxic/Irritant.

Availability
— Standard sample supplied.

Current uses
— Styrene polymers, copolymers (high impact polystyrene) and terpolymers (ABS, SAN). Used in methacrylate—butadiene—styrene (MBS) copolymer added to PVC formulations as an impact modifier. Polyesters.

Applications
— Diverse applications including packaging materials, kitchenware, household appliances, adhesives, can sealants, and cargo and bulk storage containers. Crystal polystyrene — plates, drinking cups, dessert tubs; expanded polystyrene — trays (frozen and chilled foods — fish, meats), cups (vending machines) and fast foods; high impact polystyrene — dairy products, aseptic packaging; ABS — refrigertor lining, kitchen appliances and utensils, soft margarine tubs, yoghurt pots, measuring jugs, lemon squeezers; Glass

reinforced plastic – bulk transport
(wines); thermoset polyester – microwave
cookware. Coatings.

Methods of Characterisation – IR
Mass Spectroscopy

GC Retention Index   – 729
(DB5, 3 min at 50°C, rising 20°C/min^{-1} to
300°C, hold for 20 min.)

Purity   – 99.6%

Analytical methods   – Headspace GC (FID or selected ion
monitoring mass spectrometry) for aqueous
simulants, olive oil and foods. Azeotropic
distillation with methanol then headspace
GC or HPLC (UV detection).

References   – Draft CEN method (Fraunhofer, D).
J. Chromatography, 1981, 205, 434–437.
J. Assoc. Off. Anal. Chem., 1983, 66,
1067–1073.
Var Foda, 31, 155–160

Styrene

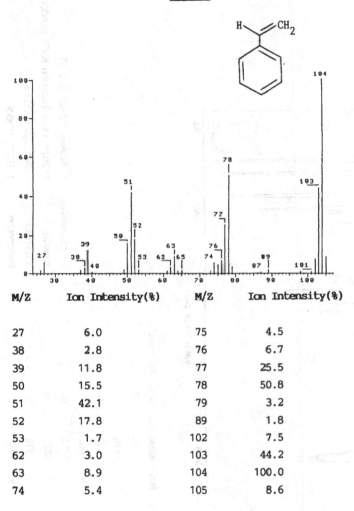

M/Z	Ion Intensity(%)	M/Z	Ion Intensity(%)
27	6.0	75	4.5
38	2.8	76	6.7
39	11.8	77	25.5
50	15.5	78	50.8
51	42.1	79	3.2
52	17.8	89	1.8
53	1.7	102	7.5
62	3.0	103	44.2
63	8.9	104	100.0
74	5.4	105	8.6

Spectrometer   :Finnigan Mat SSQ 70
Inlet System   :Capillary GC/MS
Source Temperature:150°C
Electron Energy   :70 eV
Scan Range   :25–400

Styrene   Transmission Infra Red

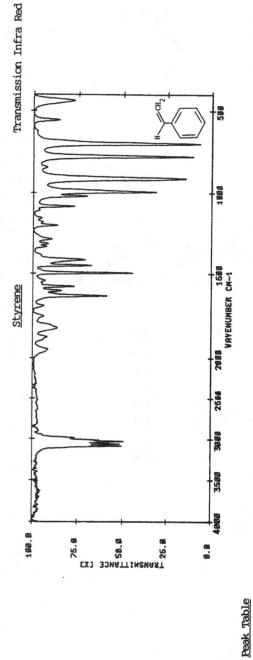

Spectrometer	: Bruker FT-IR IFS 85
Sample	: Thin film between NaCl windows
Solvents	: Not applicable
Reference	: NaCl window
Scan Time	: 1.4 sec/scan
Number of Scans: 60	

Peak Table

No.	Wave No.	Rel. Intensity	No.	Wave No.	Rel. Intensity
1	553	15	12	1493	59
2	698	100	13	1576	24
3	775	96	14	1630	44
4	908	91	15	1689	7
5	991	73	16	1750	6
6	1020	32	17	1821	15
7	1082	24	18	1876	8
8	1201	14	19	1946	9
9	1289	11	20	3028	55
10	1414	31	21	3060	52
11	1448	35	22	3083	53

Succinic acid

HOOC-CH$_2$-CH$_2$-COOH

CAS No.   – 00110–15–6
PM Ref. No.   – 24820
Restrictions   – none
Formula   – C$_4$ H$_6$ O$_4$
Molecular weight – 118.09
Alternative names– Butanedioic acid.

Physical Characteristics – White crystals, mp 185–187°C, bp 235°C. Soluble in alcohol, ether and acetone.

Handling   – Store at room temperature (25°C).
Safety   – Irritant.
Availability – Standard sample supplied.

Current uses – Colour stabiliser for PVC and acrylonitrile resins. Crosslinking agent for acrylic films. Polymer with ethylene glycol.
Applications – Lacquers. Heat curable, one package polyurethane coatings.

Methods of Characterisation – IR
              Mass Spectroscopy

Purity   – 99%

Succinic acid

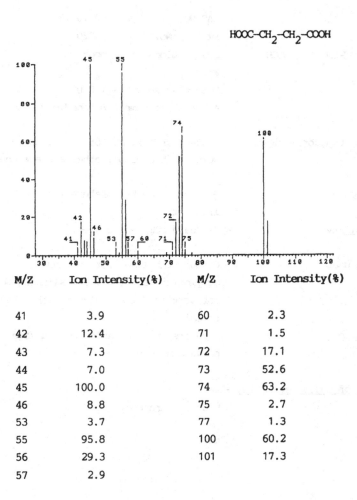

HOOC–CH$_2$–CH$_2$–COOH

M/Z	Ion Intensity(%)	M/Z	Ion Intensity(%)
41	3.9	60	2.3
42	12.4	71	1.5
43	7.3	72	17.1
44	7.0	73	52.6
45	100.0	74	63.2
46	8.8	75	2.7
53	3.7	77	1.3
55	95.8	100	60.2
56	29.3	101	17.3
57	2.9		

Spectrometer   :Finnigan Mat SSQ 70
Inlet System   :Capillary GC/MS
Source Temperature:150°C
Electron Energy   :70 eV
Scan Range   :25–400

Transmission Infra Red

Succinic acid

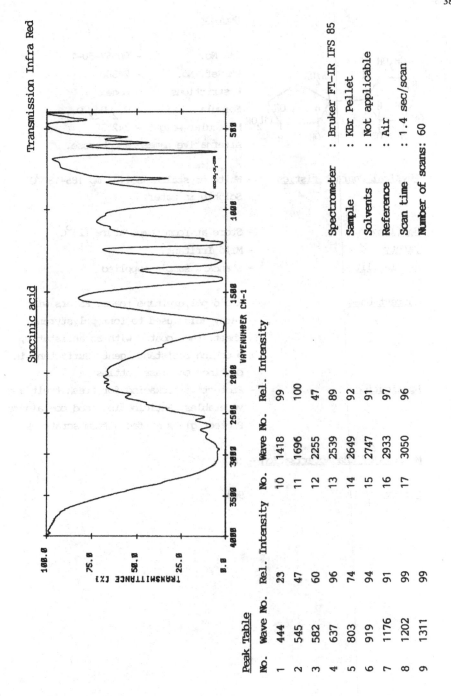

Spectrometer	: Bruker FT-IR IFS 85
Sample	: KBr Pellet
Solvents	: Not applicable
Reference	: Air
Scan time	: 1.4 sec/scan
Number of scans: 60	

Peak Table

No.	Wave No.	Rel. Intensity	No.	Wave No.	Rel. Intensity
1	444	23	10	1418	99
2	545	47	11	1696	100
3	582	60	12	2255	47
4	637	96	13	2539	89
5	803	74	14	2649	92
6	919	94	15	2747	91
7	1176	91	16	2933	97
8	1202	99	17	3050	96
9	1311	99			

Sucrose

CAS No.	– 00057–50–1
PM Ref. No.	– 24880
Restrictions	– none
Formula	– $C_{12} H_{22} O_{11}$
Molecular weight	– 342.30
Alternative names	– Saccharose.

Physical Characteristics – White crystaline solid, mp 185–187°C. Soluble in water.

Handling – Store at room temperature (25°C).

Safety – Mild Irritant.

Availability – Standard sample supplied.

Current uses – Rigid polyurethane foams. Esters with fatty acids used to form polystyrene sheet, then coated with an antistatic, blocking resistant agent. Surfactant in coatings on glass bottles.

Applications – Protective packaging for fresh fruit and vegetables. Anticlouding food containers. Protect glass surfaces from scratches.

Methods of Characterisation – IR

Purity – 99%

389

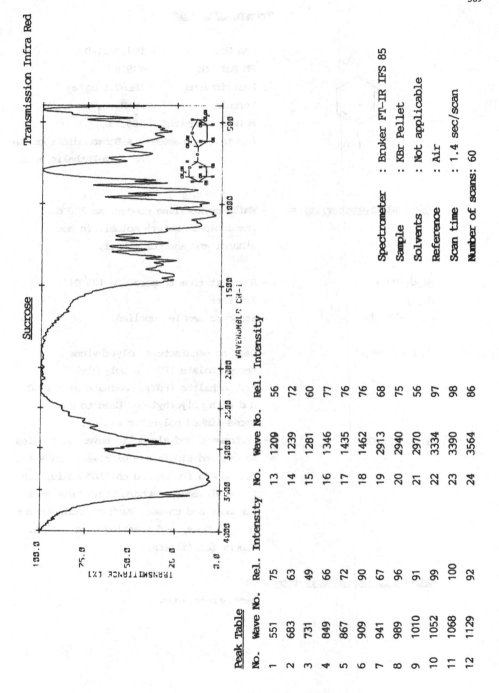

Sucrose

Transmission Infra Red

Spectrometer	: Bruker FT-IR IFS 85
Sample	: KBr Pellet
Solvents	: Not applicable
Reference	: Air
Scan time	: 1.4 sec/scan
Number of scans: 60

Peak Table

No.	Wave No.	Rel. Intensity	No.	Wave No.	Rel. Intensity
1	551	75	13	1209	56
2	683	63	14	1239	72
3	731	49	15	1281	60
4	849	66	16	1346	77
5	867	72	17	1435	76
6	909	90	18	1462	76
7	941	67	19	2913	68
8	989	96	20	2940	75
9	1010	91	21	2970	56
10	1052	99	22	3334	97
11	1068	100	23	3390	98
12	1129	92	24	3564	86

Terephthalic acid

CAS No.	– 00100-21-0
PM Ref. No.	– 24910
Restrictions	– SML=7.5 mg/kg
Formula	– $C_8 H_6 O_4$
Molecular weight	– 166.13
Alternative names	– 1,4-Benzenedicarboxylic acid, p-Phthalic acid.

Physical Characteristics — White crystalline powder, mp 300°C. Insoluble in water, soluble in hot ethanol and aqueous alkali.

Handling — Store at room temperature (25°C).

Safety — Irritant.

Availability — Standard sample supplied.

Current uses — Used to manufacture polyethylene terephthalate (PET) & polybutyl terephthalate (PBT). Laminate with PVdC, and with polyethylene. Used to make cross-linked polyester resins.

Applications — Carbonated and alcoholic beverage bottles. Re-use and single use microwave trays for pre-cooked frozen and chilled meals. Lid for modified atmosphere trays. Wrappers for meat and cheese. Part of laminate for boil-in-bags, and roasting bags. Films. Fibers for filters.

Methods of Characterisation — IR
Mass Spectroscopy

| GC Retention Index | – 1905 |
| | (DB5, 3 min at 50°C, rising 20°C/min⁻¹ to 300°C, hold for 20 min.) |

GC Retention Index — – 1905
(DB5, 3 min at 50°C, rising $20°C/min^{-1}$ to 300°C, hold for 20 min.)

Purity — – Technical grade supplied by industry. Purity not declared.

Analytical methods — – HPLC (UV detection) – direct injection of aqueous simulants, bicarbonate extraction of olive oil.

References — – Draft CEN method (PIRA, UK).
Food Additives and Contaminants, 1987, 4, 267–276.
Food Additives and Contaminants, 1988, 5, 485–492.

Terephthalic acid

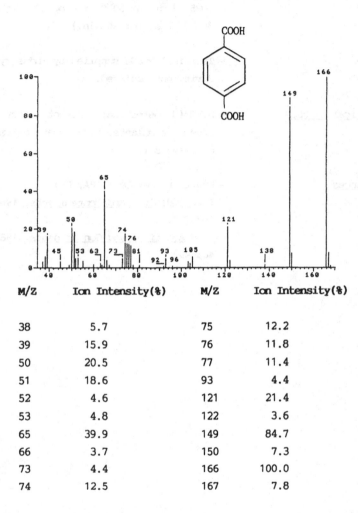

M/Z	Ion Intensity(%)	M/Z	Ion Intensity(%)
38	5.7	75	12.2
39	15.9	76	11.8
50	20.5	77	11.4
51	18.6	93	4.4
52	4.6	121	21.4
53	4.8	122	3.6
65	39.9	149	84.7
66	3.7	150	7.3
73	4.4	166	100.0
74	12.5	167	7.8

Spectrometer   :Finnigan Mat SSQ 70
Inlet System   :Capillary GC/MS
Source Temperature:150°C
Electron Energy   :70 eV
Scan Range   :25–400

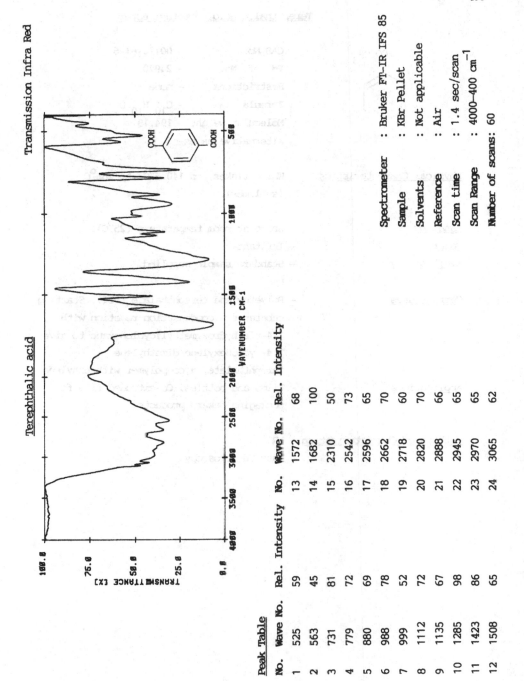

Terephthalic acid

Transmission Infra Red

Spectrometer	: Bruker FT-IR IFS 85
Sample	: KBr Pellet
Solvents	: Not applicable
Reference	: Air
Scan time	: 1.4 sec/scan
Scan Range	: 4000–400 cm^{-1}
Number of scans: 60	

Peak Table

No.	Wave No.	Rel. Intensity	No.	Wave No.	Rel. Intensity
1	525	59	13	1572	68
2	563	45	14	1682	100
3	731	81	15	2310	50
4	779	72	16	2542	73
5	880	69	17	2596	65
6	988	78	18	2662	70
7	999	52	19	2718	60
8	1112	72	20	2820	70
9	1135	67	21	2888	66
10	1285	98	22	2945	65
11	1423	86	23	2970	65
12	1508	65	24	3065	62

Terephthalic acid, dimethyl ester

COOCH$_3$

COOCH$_3$

CAS No.	– 00120-61-6
PM Ref. No.	– 24970
Restrictions	– none
Formula	– C$_{10}$ H$_{10}$ O$_4$
Molecular weight	– 194.19
Alternative names	–

Physical Characteristics — White flakes, mp 140.6°C, bp 288°C (sublimes).

Handling — Store at room temperature (25°C).

Safety — Irritant.

Availability — Standard sample supplied.

Current uses — Polyethylene terephthalate (PET). Starting substance - condensation reaction with 1,4-bis(hydroxymethyl)cyclohexane to give 1,4-cyclohexylene dimethylene terephthalate, a co-polymer with ethylene.

Applications — Films and bottles. Co-extruded film for packaging bakery products.

Methods of Characterisation — IR
Mass Spectroscopy

Purity — 99%

Terephthalic acid, dimethyl ester

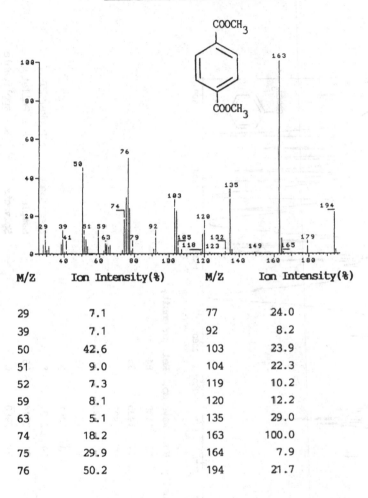

M/Z	Ion Intensity(%)	M/Z	Ion Intensity(%)
29	7.1	77	24.0
39	7.1	92	8.2
50	42.6	103	23.9
51	9.0	104	22.3
52	7.3	119	10.2
59	8.1	120	12.2
63	5.1	135	29.0
74	18.2	163	100.0
75	29.9	164	7.9
76	50.2	194	21.7

Spectrometer	:Finnigan Mat SSQ 70
Inlet System	:Capillary GC/MS
Source Temperature	:150°C
Electron Energy	:70 eV
Scan Range	:25–400

Terephthalic acid, dimethyl ester

Transmission Infra Red

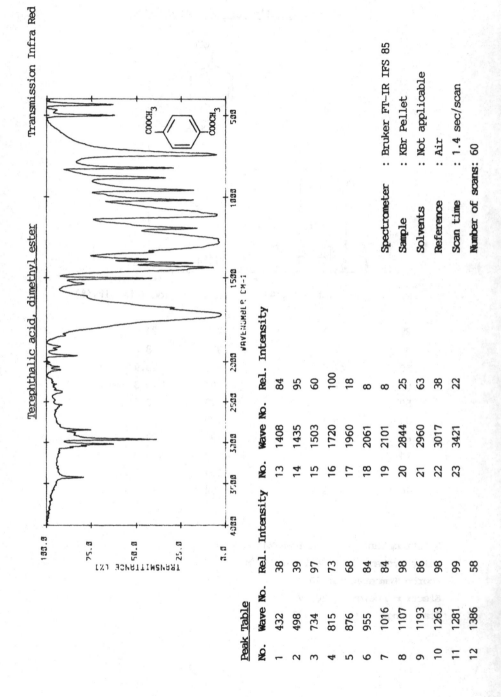

Spectrometer	: Bruker FT-IR IFS 85
Sample	: KBr Pellet
Solvents	: Not applicable
Reference	: Air
Scan time	: 1.4 sec/scan
Number of scans: 60	

Peak Table

No.	Wave No.	Rel. Intensity	No.	Wave No.	Rel. Intensity
1	432	38	13	1408	84
2	498	39	14	1435	95
3	734	97	15	1503	60
4	815	73	16	1720	100
5	876	68	17	1960	18
6	955	84	18	2061	8
7	1016	84	19	2101	8
8	1107	98	20	2844	25
9	1193	86	21	2960	63
10	1263	98	22	3017	38
11	1281	99	23	3421	22
12	1386	58			

Tetrahydrofuran

CAS No.   – 00109-99-9
PM Ref. No.   – 25150
Restrictions   – SML= 0.6mg/kg
Formula   – $C_4 H_8 O$
Molecular weight   – 72.11
Alternative names – Diethylene oxide.

Physical Characteristics	– Liquid, mp -108.5°C, bp 65.5-66.5°C. Miscible with water, alcohols and ketones.
Handling	– Store at room temperature 25°C.
Safety	– Flammable/Irritant.
Availability	– Standard sample supplied.
Current uses	– Polytetrahydrofuran. Used in the manufacture of polyurethanes and polyvinylidene chloride.
Applications	– Elastomers with fungal resistance. Adhesives/solvents for PVC. Coating cellophane.
Methods of Characterisation	– IR Mass Spectroscopy
Purity	– 99%
Analytical methods	– Analysis of aqueous and olive oil simulants by capillary headspace GC with FID (or GC/MS).
References	– Method under development (Inst. Superiore di Sanita, Rome, I).

398

Tetrahydrofuran

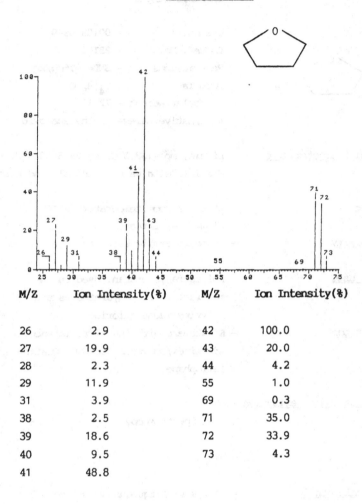

M/Z	Ion Intensity(%)	M/Z	Ion Intensity(%)
26	2.9	42	100.0
27	19.9	43	20.0
28	2.3	44	4.2
29	11.9	55	1.0
31	3.9	69	0.3
38	2.5	71	35.0
39	18.6	72	33.9
40	9.5	73	4.3
41	48.8		

Spectrometer   :Finnigan Mat SSQ 70
Inlet System   :Capillary GC/MS
Source Temperature:150°C
Electron Energy   :70 eV
Scan Range   :25-400

Tetrahydrofuran

Transmission Infra Red

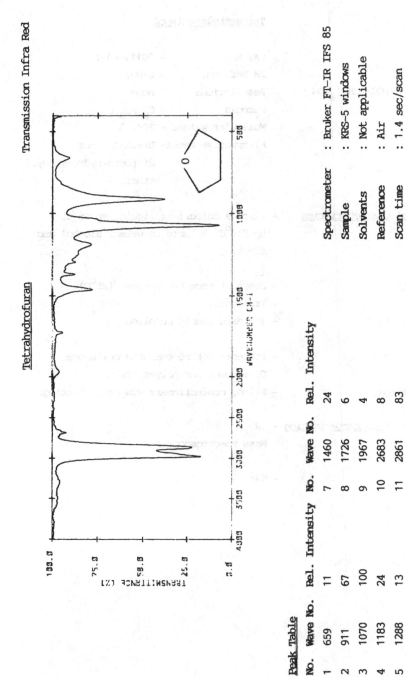

Spectrometer	: Bruker FT-IR IFS 85
Sample	: KRS-5 windows
Solvents	: Not applicable
Reference	: Air
Scan time	: 1.4 sec/scan
Number of scans: 60	

Peak Table

No.	Wave No.	Rel. Intensity	No.	Wave No.	Rel. Intensity
1	659	11	7	1460	24
2	911	67	8	1726	6
3	1070	100	9	1967	4
4	1183	24	10	2683	8
5	1288	13	11	2861	83
6	1364	13	12	2976	89

Tetraethyleneglycol

H-[OCH$_2$CH$_2$]$_4$-OH

CAS No.	– 00112-60-7
PM Ref. No.	– 25090
Restrictions	– none
Formula	– C$_8$ H$_{18}$ O$_5$
Molecular weight	– 194.23
Alternative names	– Diethyl ether di-[beta-hydroxyethyl] ether.

Physical Characteristics — Viscous colourless liquid, mp -6°C, bp 314°C. Soluble in water, alcohol and ether.

Handling — Store at room temperature (25°C).

Safety — Irritant.

Availability — Standard sample supplied.

Current uses — Softener for regenerated cellulose. Co-monomer for polyesters.

Applications — Films, confectionary wrappers. Coatings.

Methods of Characterisation — IR
Mass Spectroscopy

Purity — 99%

Tetraethyleneglycol

$$H-[OCH_2CH_2]_4-OH$$

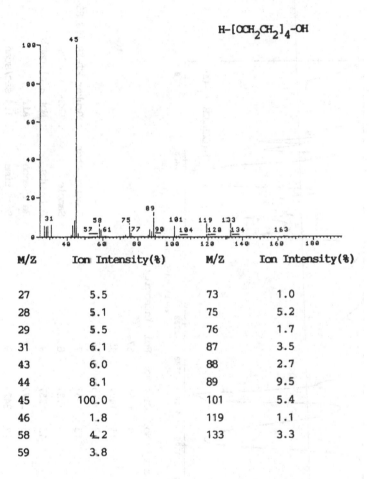

M/Z	Ion Intensity(%)	M/Z	Ion Intensity(%)
27	5.5	73	1.0
28	5.1	75	5.2
29	5.5	76	1.7
31	6.1	87	3.5
43	6.0	88	2.7
44	8.1	89	9.5
45	100.0	101	5.4
46	1.8	119	1.1
58	4.2	133	3.3
59	3.8		

Spectrometer   :Finnigan Mat SSQ 70
Inlet System   :Capillary GC/MS
Source Temperature:150°C
Electron Energy   :70 eV
Scan Range   :25–400

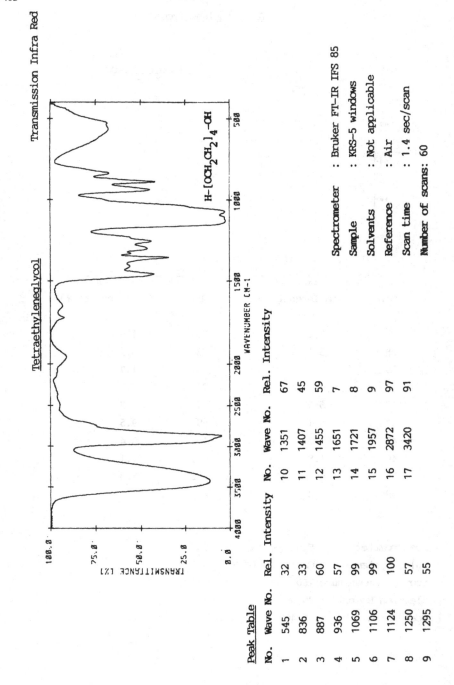

Tetraethyleneglycol

Transmission Infra Red

H-[OCH$_2$CH$_2$]$_4$-OH

Peak Table

No.	Wave No.	Rel. Intensity	No.	Wave No.	Rel. Intensity
1	545	32	10	1351	67
2	836	33	11	1407	45
3	887	60	12	1455	59
4	936	57	13	1651	7
5	1069	99	14	1721	8
6	1106	99	15	1957	9
7	1124	100	16	2872	97
8	1250	57	17	3420	91
9	1295	55			

Spectrometer	:	Bruker FT-IR IFS 85
Sample	:	KRS-5 windows
Solvents	:	Not applicable
Reference	:	Air
Scan time	:	1.4 sec/scan
Number of scans:	60	

N.N.N 'N'-Tetrakis(2-Hydroxypropyl)Ethylenediamine

```
        OH                  OH        CAS No.            - 00102-60-3
   CH3-CH-CH2          CH2-CH-CH3     PM Ref. No.        - 25180
        \               /
         N-CH2-CH2-N               Restrictions       - none
        /               \
   CH3-CH-CH2          CH2-CH-CH3     Formula            - C14 H32 N2 O4
        OH                  OH        Molecular weight - 292.42
```

$C_{14} H_{32} N_2 O_4$

Alternative names- Entprol,
(ethylenedinitrilo)
tetra-2-propanol.

Physical Characteristics   - Clear viscous liquid, bp 175-181°C
                              /0.001 bar, miscible with water, soluble
                              in methanol. Unstable, decomposes under GC
                              conditions.

Safety   - Irritant.
Availability   - No sample supplied.

Current uses   - Used as cross linking agent for
                              polyurethane foams, and in polyamide
                              resin.
Applications   - Coatings.

2,4-Toluene diisocyanate

CAS No.	– 00584-84-9
PM Ref. No.	– 25210
Restrictions	– QM(T)= 1mg/kg in FP (expressed as NCO).
Formula	– $C_9 H_6 N_2 O_2$
Molecular weight	– 174.16
Alternative names	– 4-Methyl-1,3-phenylene diisocyanate.

Physical Characteristics – Liquid, mp 19.5-21.5°C, bp 115-120°C/10mm. Soluble in most organic solvents.

Handling – Store at room temperature (25°C). Light sensitive.

Safety – Poison/Irritant.

Availability – Standard sample supplied.

Current uses – Used in the manufacture of polyurethanes.

Applications – Polyurethane tubing for food manufacturing applications. Used to make adhesives in seals for thin films, in polyester paperboard laminates (e.g susceptors) and in multi-layer high barrier plastics (e.g shelf stables) and in 'boil-in-the-bag' laminates.

Methods of Characterisation – IR
Mass Spectroscopy

Purity – 80% (2,6-Toluene diisocyanate impurity).

Analytical methods — Isocyanates in materials and articles are analysed by solvent extraction with ethanol in toluene with concurrent urethane derivative formation, clean-up by liquid/liquid partition and solid phase cartridge chromatography and determination by capillary GC with nitrogen selective detection. Phenyl isocyanate and 1,4-butanediisocyanate are used as internal standards.

References — Draft CEN Method (MAFF, FScL. Norwich, UK).

2,4-Toluene diisocyanate

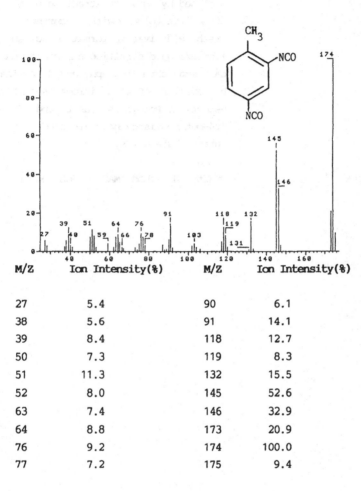

M/Z	Ion Intensity(%)	M/Z	Ion Intensity(%)
27	5.4	90	6.1
38	5.6	91	14.1
39	8.4	118	12.7
50	7.3	119	8.3
51	11.3	132	15.5
52	8.0	145	52.6
63	7.4	146	32.9
64	8.8	173	20.9
76	9.2	174	100.0
77	7.2	175	9.4

Spectrometer   :Finnigan Mat SSQ 70
Inlet System   :Capillary GC/MS
Source Temperature:150°C
Electron Energy   :70 eV
Scan Range   :25-400

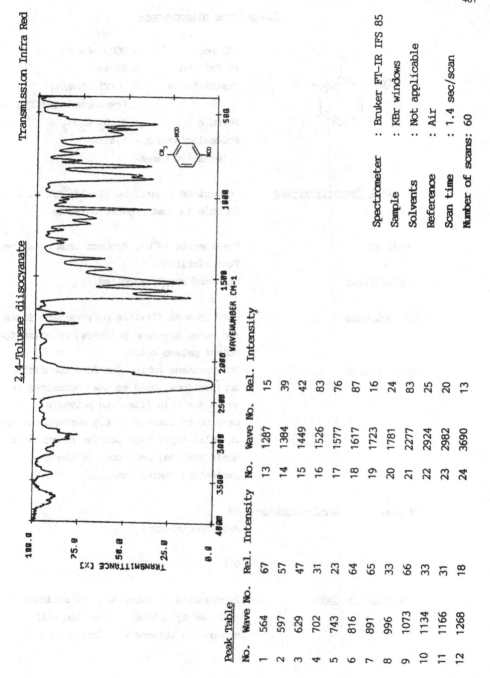

2,4-Toluene diisocyanate — Transmission Infra Red

Spectrometer : Bruker FT-IR IFS 85
Sample   : KBr windows
Solvents   : Not applicable
Reference   : Air
Scan time   : 1.4 sec/scan
Number of scans: 60

Peak Table

No.	Wave No.	Rel. Intensity	No.	Wave No.	Rel. Intensity
1	564	67	13	1287	15
2	597	57	14	1384	39
3	629	47	15	1449	42
4	702	31	16	1526	83
5	743	23	17	1577	76
6	816	64	18	1617	87
7	891	65	19	1723	16
8	996	33	20	1781	24
9	1073	66	21	2277	83
10	1134	33	22	2924	25
11	1166	31	23	2982	20
12	1268	18	24	3690	13

407

2,6-Toluene diisocyanate

CAS No.	– 00091-08-7
PM Ref. No.	– 25240
Restrictions	– QM(T)= 1mg/kg (expressed as NCO).
Formula	– $C_9 H_6 N_2 O_2$
Molecular weight	– 174.16
Alternative names-	

Physical Characteristics – Colourless liquid, bp 129-133°C/0.02 bar. Soluble in most organic solvents.

Handling – Refrigerate (4°C). Protect from moisture.

Safety – Toxic/Irritant.

Availability – Standard sample supplied.

Current uses – Used to make flexible polyurethane foams and other urethane polymers. Catalyst for polypropylene oxide.

Applications – Polyurethane tubing for food manufacturing applications. Used to make adhesives in seals for thin films, in polyester paperboard laminates (e.g susceptors) and in multi-layer high barrier plastics (e.g shelf stables) and 'boil-in-the-bag' laminates. Baking enamels.

Methods of Characterisation – IR
Mass Spectroscopy

Purity – 97%

Analytical methods – Isocyanates in materials and articles are analysed by solvent extraction with ethanol in toluene with concurrent

urethane derivative formation, clean-up by
liquid/liquid partition and solid phase
cartridge chromatography and determination
by capillary GC with nitrogen selective
detection. Phenyl isocyanate and
1,4-butanediisocyanate are used as
internal standards.

References   - Draft CEN Method (MAFF, FScL. Norwich,
UK).

2,6-Toluene diisocyanate

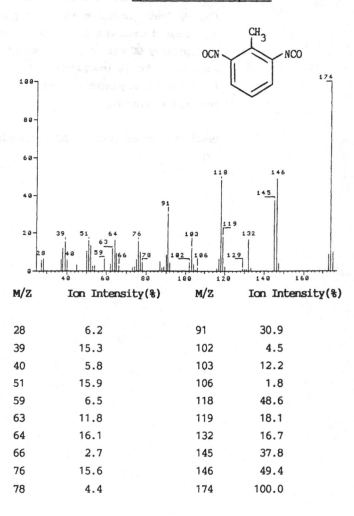

M/Z	Ion Intensity(%)	M/Z	Ion Intensity(%)
28	6.2	91	30.9
39	15.3	102	4.5
40	5.8	103	12.2
51	15.9	106	1.8
59	6.5	118	48.6
63	11.8	119	18.1
64	16.1	132	16.7
66	2.7	145	37.8
76	15.6	146	49.4
78	4.4	174	100.0

Spectrometer   :Finnigan Mat SSQ 70
Inlet System   :Capillary GC/MS
Source Temperature:150°C
Electron Energy   :70 eV
Scan Range   :25-400

411

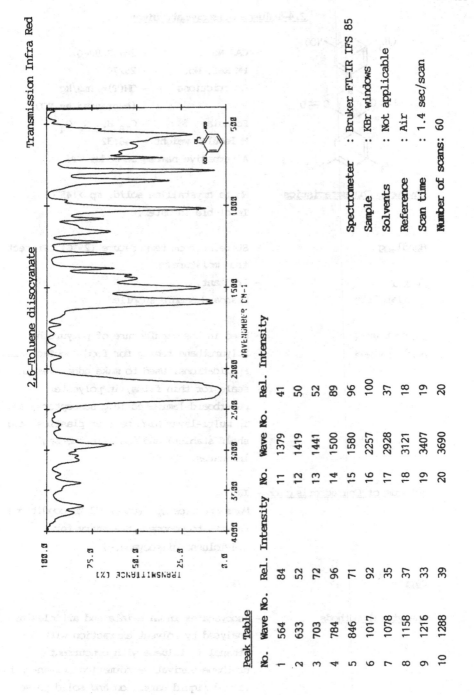

2,6-Toluene diisocyanate

Transmission Infra Red

Peak Table

No.	Wave No.	Rel. Intensity	No.	Wave No.	Rel. Intensity
1	564	84	11	1379	41
2	633	52	12	1419	50
3	703	72	13	1441	52
4	784	96	14	1500	89
5	846	71	15	1580	96
6	1017	92	16	2257	100
7	1078	35	17	2928	37
8	1158	37	18	3121	18
9	1216	33	19	3407	19
10	1288	39	20	3690	20

Spectrometer : Bruker FT-IR IFS 85
Sample : KBr windows
Solvents : Not applicable
Reference : Air
Scan time : 1.4 sec/scan
Number of scans: 60

2,4-Toluene diisocyanate dimer

CAS No.	– 26747-90-0
PM Ref. No.	– 25270
Restrictions	– QM(T)= 1mg/kg
	(expressed as NCO).
Formula	– $C_{18} H_{12} N_4 O_4$
Molecular weight	– 348.32
Alternative names	– Desmodur TT.

Physical Characteristics   – White crystalline solid, mp >145°C.
Insoluble in water.

Handling   – Store at room temperature (25°C). Protect
from moisture.

Safety   – Irritant.

Availability   – Standard sample supplied.

Current uses   – Used in the manufacture of polyurethanes.

Applications   – Polyurethane tubing for food manufacturing
applications. Used to make adhesives in
seals for thin films, in polyester
paperboard laminates (e.g susceptors) and
in multi-layer high barrier plastics (e.g
shelf stables) and 'boil-in-the-bag'
laminates.

Methods of Characterisation – IR
Mass Spectroscopy (under GC/MS conditions
reverts to monomer. See entry for
2,4-Toluene diisocyanate).

Purity   – 99%

Analytical methods   – Isocyanates in materials and articles are
analysed by solvent extraction with
ethanol in toluene with concurrent
urethane derivative formation, clean-up by
liquid/liquid partition and solid phase
cartridge chromatography and determination

by capillary GC with nitrogen selective
detection. Phenyl isocyanate and
1,4-butanediisocyanate are used as
internal standards.

References   - Draft CEN Method (MAFF, FScL. Norwich,
UK).

414

2,4-Toluene diisocyanate dimer   Transmission Infra Red

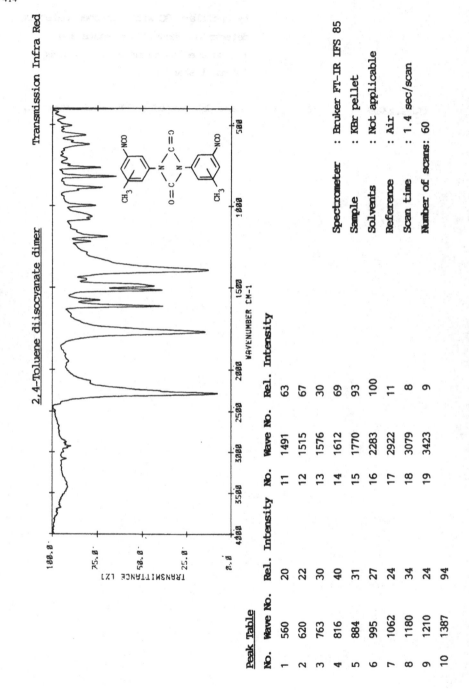

Spectrometer : Bruker FT-IR IFS 85
Sample : KBr pellet
Solvents : Not applicable
Reference : Air
Scan time : 1.4 sec/scan
Number of scans: 60

Peak Table

No.	Wave No.	Rel. Intensity	No.	Wave No.	Rel. Intensity
1	560	20	11	1491	63
2	620	22	12	1515	67
3	763	30	13	1576	30
4	816	40	14	1612	69
5	884	31	15	1770	93
6	995	27	16	2283	100
7	1062	24	17	2922	11
8	1180	34	18	3079	8
9	1210	24	19	3423	9
10	1387	94			

2,4,6-Triamino-1,3,5-Triazine

CAS No.	— 00108—78—1
PM Ref. No.	— 25420
Restrictions	— SML=30 mg/kg
Formula	— $C_3 H_6 N_6$
Molecular weight	— 126.12

Alternative names— Melamine, Cyanuramide.

Physical Characteristics — White powder, mp 354°C, slightly soluble in water. Soluble in hot ethanol.

Handling — Store at room temperature (25°C).
Safety — Irritant.
Availability — Standard sample supplied.

Current uses — Synthesis of melamine—formaldehyde resins.
Applications — Used to make picnic-ware, kitchen worktops, and to add strength to paper. Coatings.

Methods of Characterisation — IR
Mass Spectroscopy

GC Retention Index — 1640
(DB5, 3 min at 50°C, rising 20°C/min^{-1} to 300°C, hold for 20 min.)

Purity — 99.8%

Analytical methods — HPLC analysis using phosphate buffered mobile phase (ODS column) and UV detection (235 nm). Aqueous simulants directly injected, olive oil extracted with water and extract injected directly.

References

— Method under development (TNO—Inst, Zeist, NL).

Food Addit. Contam., 1990, 7, 21—27.

2,4,6-Triamino-1,3,5-Triazine

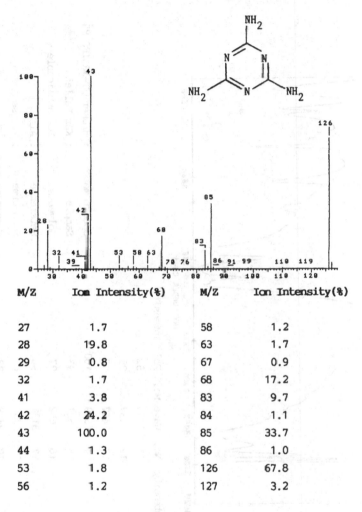

M/Z	Ion Intensity(%)	M/Z	Ion Intensity(%)
27	1.7	58	1.2
28	19.8	63	1.7
29	0.8	67	0.9
32	1.7	68	17.2
41	3.8	83	9.7
42	24.2	84	1.1
43	100.0	85	33.7
44	1.3	86	1.0
53	1.8	126	67.8
56	1.2	127	3.2

Spectrometer   :Finnigan Mat SSQ 70
Inlet System   :Capillary GC/MS
Source Temperature:150°C
Electron Energy   :70 eV
Scan Range   :25-400

418

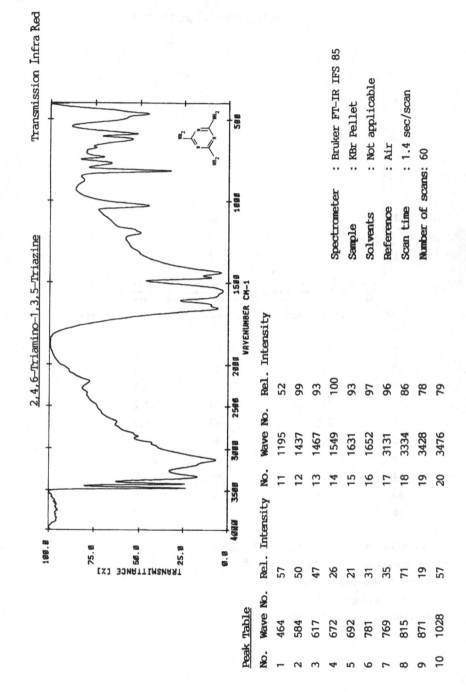

Transmission Infra Red

2,4,6-Triamino-1,3,5-Triazine

Spectrometer : Bruker FT-IR IFS 85
Sample : KBr Pellet
Solvents : Not applicable
Reference : Air
Scan time : 1.4 sec/scan
Number of scans: 60

Peak Table

No.	Wave No.	Rel. Intensity	No.	Wave No.	Rel. Intensity
1	464	57	11	1195	52
2	584	50	12	1437	99
3	617	47	13	1467	93
4	672	26	14	1549	100
5	692	21	15	1631	93
6	781	31	16	1652	97
7	769	35	17	3131	96
8	815	71	18	3334	86
9	871	19	19	3428	78
10	1028	57	20	3476	79

Triethyleneglycol

OH–(CH$_2$–CH$_2$–O)$_2$–CH$_2$CH$_2$OH

CAS No.	– 00112–27–6
PM Ref. No.	– 25510
Restrictions	– none
Formula	– C$_6$ H$_{14}$ O$_4$
Molecular weight	– 150.18
Alternative names–	

Physical Characteristics — Colourless liquid, mp –7°C, bp 285°C.
Soluble in water, alcohol and benzene.

Handling — Store at room temperature (25°C).
Hygroscopic.

Safety — Irritant

Availability — Standard sample supplied.

Current uses — Polyester resins.

Applications — Coatings.

Methods of Characterisation — IR
Mass Spectroscopy

Purity — 99%

Triethyleneglycol

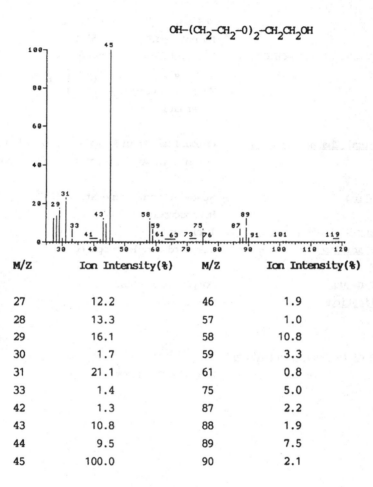

OH—(CH$_2$—CH$_2$—O)$_2$—CH$_2$CH$_2$OH

M/Z	Ion Intensity(%)	M/Z	Ion Intensity(%)
27	12.2	46	1.9
28	13.3	57	1.0
29	16.1	58	10.8
30	1.7	59	3.3
31	21.1	61	0.8
33	1.4	75	5.0
42	1.3	87	2.2
43	10.8	88	1.9
44	9.5	89	7.5
45	100.0	90	2.1

Spectrometer   :Finnigan Mat SSQ 70
Inlet System   :Capillary GC/MS
Source Temperature:150°C
Electron Energy   :70 eV
Scan Range   :25–400

421

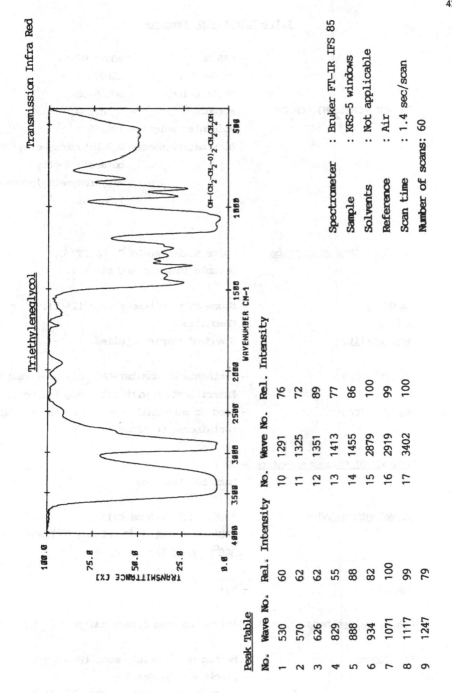

Triethyleneglycol

Transmission Infra Red

OH-(CH_2-CH_2-O)_2-CH_2CH_2OH

Spectrometer	: Bruker FT-IR IFS 85
Sample	: KRS-5 windows
Solvents	: Not applicable
Reference	: Air
Scan time	: 1.4 sec/scan
Number of scans: 60	

Peak Table

No.	Wave No.	Rel. Intensity	No.	Wave No.	Rel. Intensity
1	530	60	10	1291	76
2	570	62	11	1325	72
3	626	62	12	1351	89
4	829	55	13	1413	77
5	888	88	14	1455	86
6	934	82	15	2879	100
7	1071	100	16	2919	99
8	1117	99	17	3402	100
9	1247	79			

1,1,1-Trimethylol Propane

$$CH_3-CH_2-C(CH_2OH)_2-CH_2OH$$

CAS No.	– 00077–99–6
PM Ref. No.	– 25600
Restrictions	– SML=6 mg/kg
Formula	– $C_6H_{14}O_3$
Molecular weight	– 134.18
Alternative names	– 2,2–Dihydroxymethyl–1–butanol; 2–ethyl 2–hydroxymethylpropane–1,3–diol.

Physical Characteristics — White solid, mp 59°C, bp 297°C. Soluble in water and alcohol.

Handling — Store at room temperature (25°C).
Safety — Combustible.
Availability — Standard sample supplied.

Current uses — Synthesis of unsaturated polyester resins. Esterification with oils, e.g tall oil.

Applications — Used to make dual ovenable trays. Storage containers. Coatings.

Methods of Characterisation — IR
Mass Spectroscopy

GC Retention Index — 1305, 1329 (fragments)
(DB5, 3 min at 50°C, rising 20°C/min^{-1} to 300°C, hold for 20 min.)

Purity — 97%

Analytical methods — Extraction and direct analysis by GC.

References — Method under development (Packforsk, Stockholm, Sweden.)
J. Chromatogr. Sci., 1984, 22, 10–12
Zh. Anal. Khim., 1975, 30, 391–393

1,1,1-Trimethylol propane

$$CH_3-CH_2-C(CH_2OH)_2-CH_2OH$$

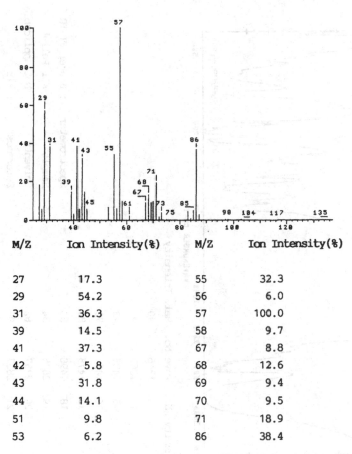

M/Z	Ion Intensity(%)	M/Z	Ion Intensity(%)
27	17.3	55	32.3
29	54.2	56	6.0
31	36.3	57	100.0
39	14.5	58	9.7
41	37.3	67	8.8
42	5.8	68	12.6
43	31.8	69	9.4
44	14.1	70	9.5
51	9.8	71	18.9
53	6.2	86	38.4

Spectrometer   :Finnigan Mat SSQ 70
Inlet System   :Capillary GC/MS
Source Temperature:150°C
Electron Energy   :70 eV
Scan Range   :25–400

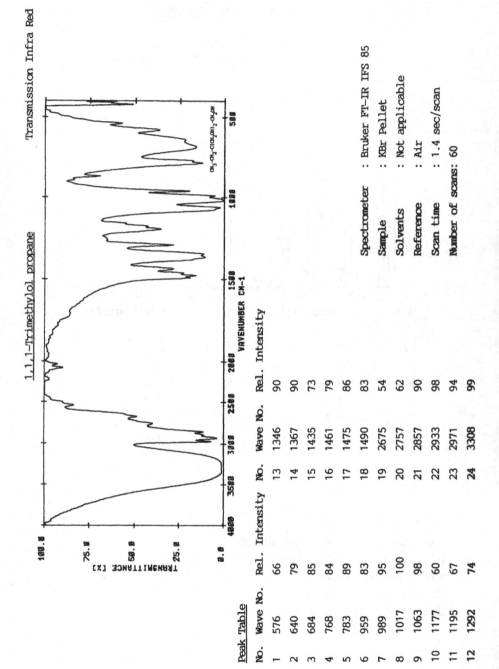

1,1,1-Trimethylol propane   Transmission Infra Red

Spectrometer	: Bruker FT-IR IFS 85
Sample	: KBr Pellet
Solvents	: Not applicable
Reference	: Air
Scan time	: 1.4 sec/scan
Number of scans: 60	

Peak Table

No.	Wave No.	Rel. Intensity	No.	Wave No.	Rel. Intensity
1	576	66	13	1346	90
2	640	79	14	1367	90
3	684	85	15	1435	73
4	768	84	16	1461	79
5	783	89	17	1475	86
6	959	83	18	1490	83
7	989	95	19	2675	54
8	1017	100	20	2757	62
9	1063	98	21	2857	90
10	1177	60	22	2933	98
11	1195	67	23	2971	94
12	1292	74	24	3308	99

Tripropyleneglycol

HO(CH$_2$)$_3$-O-(CH$_2$)$_3$-O-(CH$_2$)$_3$-OH

CAS No.	— 24800-44-0
PM Ref. No.	— 25910
Restrictions	— none
Formula	— C$_9$ H$_{20}$ O$_4$
Molecular weight	— 192.26
Alternative names—	

Physical Characteristics — Liquid, bp 273°C. Mixture of isomers.

Handling — Store at room temperature (25°C).

Safety — Irritant.

Availability — Standard sample supplied. Mixture of isomers.

Current uses — Used in inks. Used in polyester resins.

Applications — Carbonated beverage bottles, coffee makers and toasters. Lubricant for food machinery.

Methods of Characterisation — IR
Mass Spectroscopy

Purity — 97%

426

Tripropyleneglycol

$HO(CH_2)_3-O-(CH_2)_3-O-(CH_2)_3-OH$

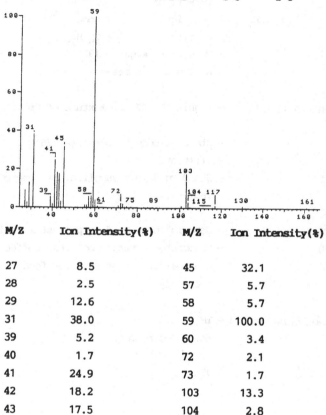

M/Z	Ion Intensity(%)	M/Z	Ion Intensity(%)
27	8.5	45	32.1
28	2.5	57	5.7
29	12.6	58	5.7
31	38.0	59	100.0
39	5.2	60	3.4
40	1.7	72	2.1
41	24.9	73	1.7
42	18.2	103	13.3
43	17.5	104	2.8
44	3.0	117	1.2

Spectrometer   :Finnigan Mat SSQ 70
Inlet System   :Capillary GC/MS
Source Temperature:150°C
Electron Energy   :70 eV
Scan Range   :25-400

Transmission Infra Red

Tripropyleneglycol

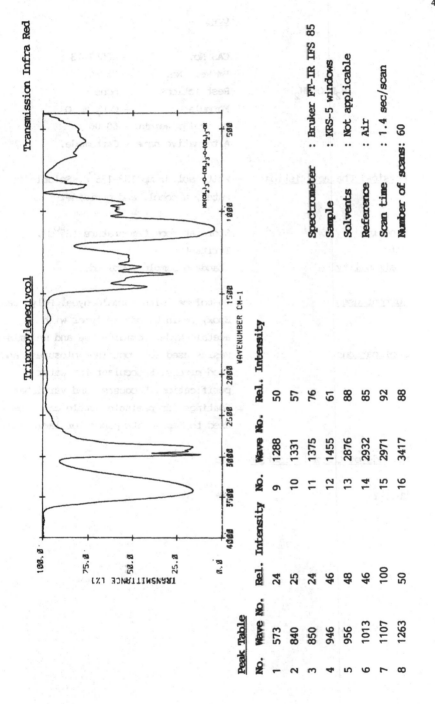

Spectrometer	: Bruker FT-IR IFS 85
Sample	: KRS-5 windows
Solvents	: Not applicable
Reference	: Air
Scan time	: 1.4 sec/scan
Number of scans: 60	

Peak Table

No.	Wave No.	Rel. Intensity	No.	Wave No.	Rel. Intensity
1	573	24	9	1288	50
2	840	25	10	1331	57
3	850	24	11	1375	76
4	946	46	12	1455	61
5	956	48	13	2876	88
6	1013	46	14	2932	85
7	1107	100	15	2971	92
8	1263	50	16	3417	88

Urea

CAS No. — 00057–13–6
PM Ref. No. — 25960
Restrictions — none
Formula — $C H_4 N_2 O$
Molecular weight — 60.06
Alternative names— Carbamide.

$$H_2N-\overset{\overset{\displaystyle O}{\|}}{C}-NH_2$$

Physical Characteristics — White solid, mp 133–135°C. Soluble in water, alcohol, and pyridine.

Handling — Store at room temperature (25°C).
Safety — Irritant.
Availability — Standard sample supplied.

Current uses — Co-polymer with formaldehyde. Polyureas. Epoxy resin blends. Polymer with acetaldehyde, formaldehyde and melamine.

Applications — Resins used for container closures, and food mixers. Flocculant for water purification. Lacquers, and varnishes. Coatings for peelable bottle cap liners. Used to impregnate paper for laminates.

Methods of Characterisation — IR

Purity — 99%

Urea

Transmission Infra Red

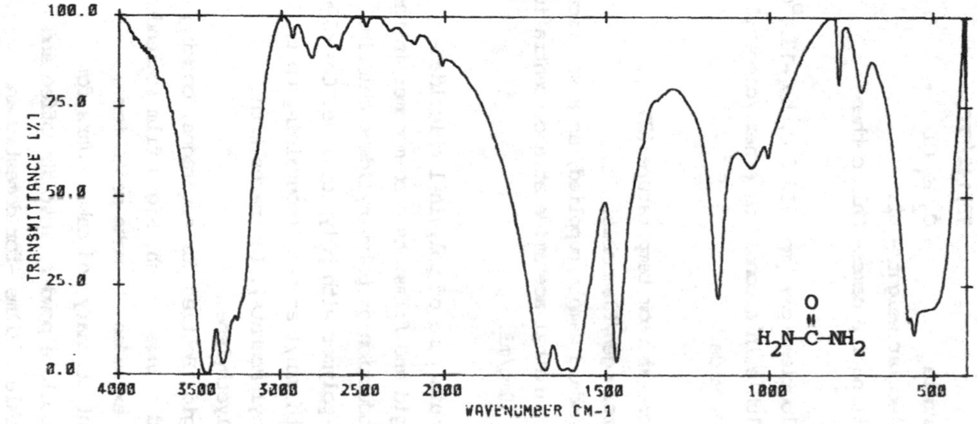

Peak Table

No.	Wave No.	Rel. Intensity	No.	Wave No.	Rel. Intensity
1	558	94	9	1682	100
2	717	25	10	2640	12
3	788	25	11	2678	11
4	1056	51	12	2804	15
5	1154	83	13	2924	8
6	1466	98	14	3344	97
7	1602	100	15	3449	100
8	1626	100			

Spectrometer : Bruker FT-IR IFS 85
Sample : KBr Pellet
Solvents : Not applicable
Reference : Air
Scan time : 1.4 sec/scan
Number of scans: 60

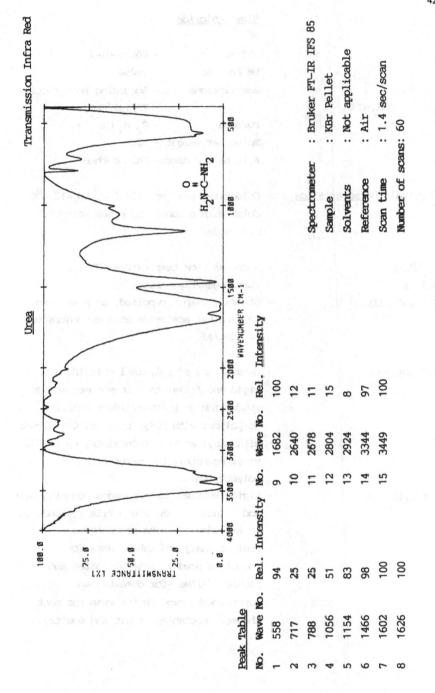

Vinyl chloride

$CH_2=CHCl$

CAS No.   – 00075-01-4
PM Ref. No.   – 26050
Restrictions   – According to Directive
                78/142/EEC.
Formula   – $C_2 H_3 Cl$
Molecular weight – 62.50
Alternative names– Chloroethene.

Physical Characteristics   – Colourless gas, mp $-153.8^{o}C$, bp $-13.4^{o}C$. Soluble in alcohol and ether, dimethyl acetamide.

Handling   – Store at room temperature ($25^{o}C$).

Safety   – Carcinogen/Flammable.

Availability   – Standard sample supplied, as a solution in dimethyl acetamide at a concentration of 10mg/ml.

Current uses   – Manufacture of polyvinyl chloride – rigid and films. As a co-monomer in the production of polyvinylidene chloride. Co-polymer with polypropylene. Co-polymer with vinyl acetate (containing up tp 20% vinyl acetate). Laminates with polyethylene.

Applications   – Rigid -Bottles for beverages, cooking oils and orange squash. Rigid film (calendered or extruded) to make nests for confectionary and cakes. Jars for chocolate powder, instant coffee and pickles. Films -for domestic and supermarket wrap. Shrink wrap for meat, cheese, vegetables, fruit and poultry.

'Cling'film. Coating for paper,
regenerated cellulose and polypropylene
films, to improve barrier properties. Can
coatings, good adhesion to steel &
aluminium. Seals for jar lids and bottle
tops. Trays for 'modified atmosphere
packaging'. Refrigerator trays.

Methods of Characterisation - Mass Spectroscopy

Purity   -

Analytical methods   - Headspace GC with flame ionisation
                              detection. Plastics are dissolved in
                              dimethylacetamide and equilibrated at 70°C
                              prior to headspace sampling. A similar
                              approach is used for foods and simulants.
                              Confirmation by GC/MS selected ion
                              monitoring.

References   - Die Angewarte. Makromol. Chemie., 1975,
                              47, 29-41.
                              J. Assoc. Off. Anal. Chem., 1978, 61,
                              813-819.
                              J. Assoc. Publ. Anal., 1981, 19, 39-49.
                              Fd. Cosmet. Toxicol., 1982, 20, 603-610.

Vinyl chloride

$$CH_2=CHCl$$

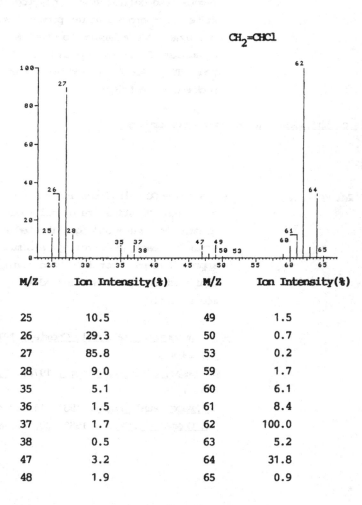

M/Z	Ion Intensity(%)	M/Z	Ion Intensity(%)
25	10.5	49	1.5
26	29.3	50	0.7
27	85.8	53	0.2
28	9.0	59	1.7
35	5.1	60	6.1
36	1.5	61	8.4
37	1.7	62	100.0
38	0.5	63	5.2
47	3.2	64	31.8
48	1.9	65	0.9

Spectrometer   :Finnigan Mat SSQ 70
Inlet System   :Capillary GC/MS
Source Temperature:150°C
Electron Energy   :70 eV
Scan Range   :25-400

Vinylidene chloride

$CH_2=CCl_2$

CAS No.   - 00075-35-4
PM Ref. No.   - 26110
Restrictions   - QM= 5mg/kg in FP,
                     SML= not detectable
                     (DL= 0.05mg/kg)
Formula   - $C_2 H_2 Cl_2$
Molecular weight - 96.94
Alternative names- 1,1-Dichloroethylene,
                          1,1-Dichloroethene.

Physical Characteristics   - Colourless liquid, mp $-122^{\circ}C$, bp $31.9^{\circ}C$.
Soluble in dimethyl acetamide and most
organic solvents. Stabilised with 0.02%
4-methoxy-phenol.

Handling   - Refrigerate $(4^{\circ}C)$. Ventillate.
Safety   - Harmful/Flammable.
Availability   - Standard sample supplied as a solution in
dimethyl acetamide at a concentration of
10mg/ml.

Current uses   - Polyvinylidene chloride (PVDC) polymers.
Co-monomer with vinyl chloride,
acrylonitrile, methacrylonitrile and
methyl acrylate/methacrylate. Laminates
with polyethylene terephthalate.
PVDC-co-extruded with low density
polyethylene.
Applications   - PVDC/PVC co-polymer used to produce a
homogeneous film for vacuum packing of
meat and poultry, and for 'chub' packs for
cheese and pate. Shrink wrap. Lacquer
resins or films to improve barrier

Re